CU00601112

PREFACE

THIS book has been written as an introduction to the Calculus. The arrangement has been adopted after considerable experience with a class of beginners, and it has been found to lead to a rapid acquisition of a working knowledge of the subject.

The differentiation and integration of the simpler standard forms are applied as early as possible to the determination of maxima and minima, of the areas and lengths of curves, of volumes of revolution, to the solution of problems in mechanics and physics and to the expansion of simple trigonometrical functions.

Great care has been taken in the selection and arrangement of the Examples (about 750 in number), which include a series of Revision Papers.

Hints for the solution of the Examples are given in some cases.

It is assumed that the student has had some experience in drawing the graphs of curves from their equations; and that he has a working knowledge of:

ALGEBRA, to the Binomial and Exponential Theorems;

TRIGONOMETRY, to the Solution of Triangles;

The elements of STATICS AND DYNAMICS.

In addition to the above, some knowledge of ALGEBRAIC GEOMETRY, particularly in connection with the straight line

TABLE OF CONSTANTS, Etc.

In working examples, the following may be used :

$\pi = 3.1416.$ $\frac{1}{\pi} = .31831.$

1 c. ft. of water weighs 62·3 lb.

1 litre .... 1 kilogram.

1 c. cm. .... 1 gram.

Base of natural (Napierian) logarithms $-\epsilon = 2\ 7183$

$\text{Log}_{10}\epsilon = 0.4343.$

$\text{Log}_e 10 = 2.3026.$

CHAPTER I.

FUNCTION. LIMITING VALUES. DIFFERENTIAL COEFFICIENTS.

Function. If $y = x^2 + 2x$, we see that

when

$x=0$	1	2	3	...
$y=0$	3	8	15	...

Here x and y are so connected that if we make any change in the value of one of them, we must make a corresponding change in the value of the other.

When two quantities are connected in such a manner, one is said to be a *function* of the other, and both quantities are called *variables*.

One is called the *independent variable*, the other the *dependent variable*.

An *independent variable* is a quantity to which we may assign any value; a *dependent variable* is a quantity whose value is determined as soon as that of some independent variable is assigned.

When we have two or more variables, we may usually take whichever we please as the independent variable.

We denote functions of x by such symbols as $f(x)$, $F(x)$, $\phi(x)$. Sometimes it is convenient to use the letters u, v, w to denote functions of x. $ax^2 + bx + c$, $\sqrt{a^2 - x^2}$, $\sin x$, $\tan^2 x$, $1 + \sin^2 x$ are all functions of x.

In dealing with any equation of the form $y = f(x)$, we usually take x as the independent variable; y, therefore, is the dependent variable.

Limiting Values. In the Geometrical Progression

$$1 + \frac{1}{2} + \frac{1}{4} + \frac{1}{8} + \dots,$$

the sum to n terms $= \dfrac{1 - \dfrac{1}{2^n}}{1 - \dfrac{1}{2}} = 2 - \dfrac{1}{2^{n-1}}.$

As the number of terms is increased, $\dfrac{1}{2^{n-1}}$ becomes smaller and smaller, *i.e.* the sum of the series continually approaches 2; and may be made to differ from 2 by as small a quantity as we please by taking n large enough.

Therefore we say that the sum of this series to infinity is 2. This is an abbreviated way of saying that the more terms we take, the more nearly does their sum approximate to 2, for the sum is never actually 2, however great n may be taken to be.

It is sometimes expressed thus:

$$\operatorname*{Lt}_{n=\infty} \left(1 + \frac{1}{2} + \frac{1}{4} + \frac{1}{8} + \dots + \frac{1}{2^{n-1}} \right) = 2.$$

Let us consider the expression $\dfrac{(1+x)^n - 1}{x}$. When we put $x = 0$, this assumes the form $\dfrac{0}{0}$ which has no meaning, and is said to be *indeterminate*.

By the Binomial Theorem, if $x < 1$,

$$(1+x)^n = 1 + nx + \frac{n\,\overline{n-1}}{\underline{2}}\,x^2 + \frac{n\,\overline{n-1}\,\overline{n-2}}{\underline{3}}\,x^3 + \dots$$

for all values of n.

$$\therefore \frac{(1+x)^n - 1}{x} = \frac{1}{x}\left[nx + \frac{n\,\overline{n-1}}{\underline{2}}\,x^2 + \frac{n\,\overline{n-1}\,\overline{n-2}}{\underline{3}}\,x^3 + \dots \right]$$

$$= n + \frac{n\,\overline{n-1}}{\underline{2}}\,x + \frac{n\,\overline{n-1}\,\overline{n-2}}{\underline{3}}\,x^2 + \dots.$$

Now if x is diminished and continually approaches zero, each of the terms of this expression *after the first* continually diminishes and ultimately would become zero.

i.e. in the limit, when $x = 0$, $\dfrac{(1+x)^n - 1}{x} = n$,

or $$\underset{x=0}{\mathrm{Lt}}\left[\frac{(1+x)^n - 1}{x}\right] = n.$$

The following are familiar cases in Trigonometry :

$$\underset{x=0}{\mathrm{Lt}}\left(\frac{\sin x}{x}\right) = 1. \qquad \underset{x=0}{\mathrm{Lt}}\left(\frac{\tan x}{x}\right) = 1. \quad \text{(See page 35.)}$$

It must be carefully remembered that this is an abbreviated way of saying that, as x is indefinitely diminished, these expressions continually approach unity, and by taking x small enough may be made to differ from unity by as small a quantity as we please.

To find the limiting value of $\dfrac{x^2 - a^2}{x - a}$ *when* $x = a$.

If we put $x = a$, this expression takes the form $\dfrac{0}{0}$, which is indeterminate. Let $x = a + h$, where h is a small quantity.

The expression $= \dfrac{(a+h)^2 - a^2}{h} = \dfrac{2ah + h^2}{h}$

$= 2a + h$, however small we take h to be.

Hence, if h is continually diminished, the expression continually approaches $2a$.

$\therefore$ the limiting value of $\dfrac{x^2 - a^2}{x - a}$ when $x = a$ is $2a$.

Here, again, it must be remembered that this is an abbreviated way of saying that as x continually approaches the value a, the expression $\dfrac{x^2 - a^2}{x - a}$ continually approaches the value $2a$, and may be made to differ from it as little as we please.

Notation. If y is any function of x, we shall take Δx, Δy to be corresponding changes, or **increments**, in x and y respectively.

Thus if $\quad y = x^2, \qquad y + \Delta y = (x + \Delta x)^2.$

$\qquad$ „ $\qquad y = \sin x, \quad y + \Delta y = \sin(x + \Delta x).$

$\qquad$ „ $\qquad y = f(x), \quad y + \Delta y = f(x + \Delta x).$

An increment may be positive or negative, but in practice we always take the increment of **x** as positive.

Differential Coefficients.

Definition. If $f(x)$ is any function of x, and $f(x + \Delta x)$ the same function of $x + \Delta x$, then

the limiting value of $\dfrac{f(x + \Delta x) - f(x)}{\Delta x}$, *when Δx is made indefinitely small, is called the* **differential coefficient** *of $f(x)$ with respect to x.*

If $y = f(x)$ and Δx, Δy are corresponding increments of x and y,

$$y + \Delta y = f(x + \Delta x).$$
$$\therefore \ \Delta y = f(x + \Delta x) - f(x).$$
$$\therefore \ \frac{\Delta y}{\Delta x} = \frac{f(x + \Delta x) - f(x)}{\Delta x}.$$

Hence, if **y** is any function of x, and Δx, Δy are corresponding increments of x and y,

the limiting value of $\dfrac{\Delta y}{\Delta x}$, *when Δx is made indefinitely small, is the* **differential coefficient** *of y with respect to x.*

The student should be familiar with both forms of the definition.

This limiting value of $\dfrac{\Delta y}{\Delta x}$ *is denoted by the symbol* $\dfrac{dy}{dx}$.

We assume in the definition that $\dfrac{\Delta y}{\Delta x}$ *has a limit.*

Derivative. A differential coefficient is sometimes called a **derivative** or a **derived function.**

EXAMPLE 1. *Find the differential coefficient of x^2 with respect to x.*

Let $y = x^2$, and Δx, Δy be corresponding increments of x and y, so that $y + \Delta y = (x + \Delta x)^2$.

By subtraction, $\Delta y = (x + \Delta x)^2 - x^2 = 2x \Delta x + (\Delta x)^2$.

$$\therefore \frac{\Delta y}{\Delta x} = 2x + \Delta x.$$

$\therefore$ proceeding to the limit (*i.e.* making Δx indefinitely small),

$$\frac{dy}{dx} = 2x.$$

EXAMPLE 2. *If $y = x^5$, find the value of $\frac{dy}{dx}$.*

Let Δx and Δy be corresponding increments of x and y, so that $\qquad y + \Delta y = (x + \Delta x)^5$. Also $y = x^5$.

$\therefore$ by subtraction,

$$\Delta y = (x + \Delta x)^5 - x^5$$

$$= x^5 + 5x^4 \Delta x + \frac{5 \cdot 4}{1 \cdot 2} x^3 (\Delta x)^2$$

$\qquad$ + terms involving higher powers of $\Delta x - x^5$.

$\qquad\qquad\qquad\qquad$ (Binomial Theorem.)

$$\therefore \frac{\Delta y}{\Delta x} = 5x^4 + 10x^3 \Delta x + \text{terms involving powers of } \Delta x.$$

$\therefore$ proceeding to the limit, $\frac{dy}{dx} = 5x^4$.

EXAMPLE 3. *If $y = 3x^2 + 5x - 4$, find the value of $\frac{dy}{dx}$.*

Let Δx and Δy be corresponding increments of x and y so that $\qquad y + \Delta y = 3(x + \Delta x)^2 + 5(x + \Delta x) - 4$.

Also $\qquad\qquad y = 3x^2 + 5x - 4$.

$\therefore$ by subtraction, $\Delta y = 6x \Delta x + 3(\Delta x)^2 + 5\Delta x$.

$$\therefore \frac{\Delta y}{\Delta x} = 6x + 5 + 3\Delta x.$$

$\therefore$ in the limit, *i.e.* when $\Delta x = 0$, $\frac{dy}{dx} = 6x + 5$.

EXAMPLE 4. *If* $y = \dfrac{1}{3t-4}$, *find the value of* $\dfrac{dy}{dt}$.

Let Δt and Δy be corresponding increments of t and y, so that

$$y + \Delta y = \frac{1}{3 \cdot (t + \Delta t) - 4},$$

$$y = \frac{1}{3t - 4}.$$

∴ by subtraction, $\quad \Delta y = \dfrac{1}{3(t + \Delta t) - 4} - \dfrac{1}{3t - 4}$

$$= \frac{-3\Delta t}{[3(t + \Delta t) - 4](3t - 4)}.$$

$$\therefore \frac{\Delta y}{\Delta t} = \frac{-3}{[3(t + \Delta t) - 4](3t - 4)}.$$

∴ in the limit, *i.e.* when $\Delta t = 0$,

$$\frac{dy}{dt} = \frac{-3}{(3t - 4)^2}.$$

N.B. $\dfrac{dy}{dt}$ is the differential coefficient of y *with respect to* t.

The differential coefficient of a constant is zero.

Let $y = a$, where a is any constant.

$\quad\quad\quad\quad$ a cannot vary, ∴ y cannot vary ;

$$\therefore \Delta y = 0.$$

Hence, whatever value Δx may have, $\dfrac{\Delta y}{\Delta x} = 0.$

$$\therefore \frac{dy}{dx} = 0.$$

This will perhaps be more easily seen when the student has learnt that a differential coefficient is a rate-measurer, and that $\dfrac{dy}{dx}$ represents the slope of a tangent to the graph of $y = f(x)$.

If θ is the radian measure of an angle, the limiting value of the
ratio $\frac{\sin\theta}{\theta}$, where θ is made indefinitely small, is unity.

In the figure, AOB is an acute angle θ, AB an arc of a circle
whose centre is O, BN and AT are both drawn perpendicular
to OA.

The $\triangle$ OAT > the sector OAB > the $\triangle$ OAB;

i.e. $\tfrac{1}{2}$AT . OA > $\tfrac{1}{2}$OA^2 . θ > $\tfrac{1}{2}$BN . OA

$$\therefore \frac{AT}{AO} > \theta > \frac{BN}{AO} ;$$

i.e. $\tan\theta > \theta > \sin\theta.$

$$\therefore \frac{\tan\theta}{\sin\theta} > \frac{\theta}{\sin\theta} > 1.$$

Thus the value of $\frac{\theta}{\sin\theta}$ lies between the value of $\frac{\tan\theta}{\sin\theta}$ and
unity.

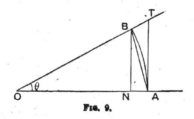

Fɪɢ. 9.

But in the limit, *i.e.* when $\theta = 0$, $\frac{\tan\theta}{\sin\theta} = \frac{1}{\cos\theta} = 1$;

and hence, since $\frac{\theta}{\sin\theta}$ lies between $\frac{\tan\theta}{\sin\theta}$ and unity,

the value of $\frac{\theta}{\sin\theta}$ is unity in the limit when $\theta = 0$.

This is only a short way of saying that, as we diminish θ
indefinitely, the value of $\frac{\theta}{\sin\theta}$ continually approaches unity.

It must be remembered that in the Calculus we usually
suppose angles to be measured in *radians*.

To prove that $\dfrac{d(\sin x)}{dx} = \cos x.$

Let $y = \sin x$, and Δx, Δy be corresponding increments of x and y, so that

$$y + \Delta y = \sin(x + \Delta x).$$

$$y = \sin x \,;$$

$\therefore$ by subtraction, $\Delta y = \sin(x + \Delta x) - \sin x$

$$= 2 \cos\left(x + \frac{\Delta x}{2}\right) \sin \frac{\Delta x}{2}.$$

$$\therefore \frac{\Delta y}{\Delta x} = \cos\left(x + \frac{\Delta x}{2}\right) \frac{\sin \dfrac{\Delta x}{2}}{\dfrac{\Delta x}{2}}.$$

But in the limit, when $\Delta x = 0$, $\dfrac{\sin \dfrac{\Delta x}{2}}{\dfrac{\Delta x}{2}} = 1.$

$$\therefore \frac{dy}{dx} = \cos x.$$

We must remember in the above that the angle x is expressed in radians.

If $y = \sin x°$, let x degrees $= \theta$ radians, so that $\dfrac{x}{180} = \dfrac{\theta}{\pi}.$

Then $\theta = \dfrac{\pi x}{180}\,; \quad \therefore \dfrac{d\theta}{dx} = \dfrac{\pi}{180}.$

Also, $y = \sin \theta \,; \quad \therefore \dfrac{dy}{dx} = \dfrac{dy}{d\theta} \times \dfrac{d\theta}{dx} = \cos\theta \times \dfrac{\pi}{180} = \dfrac{\pi}{180}\cos x°.$

$$\frac{d(\cos x)}{dx} = -\sin x.$$

The proof of this is left to the student.

To prove that $\dfrac{d(\tan x)}{dx} = \sec^2 x.$

With the same notation, $y + \Delta y = \tan(x + \Delta x)$, and $y = \tan x.$

$$\therefore \ \Delta y = \tan(x + \Delta x) - \tan x = \frac{\sin(x + \Delta x)}{\cos(x + \Delta x)} - \frac{\sin x}{\cos x}$$

$$= \frac{\sin\overline{(x + \Delta x - x)}}{\cos(x + \Delta x)\cos x}.$$

$$\therefore \ \frac{\Delta y}{\Delta x} = \frac{\dfrac{\sin \Delta x}{\Delta x}}{\cos(x + \Delta x)\cos x}.$$

$\therefore$ proceeding to the limit, $\dfrac{dy}{dx} = \dfrac{1}{\cos^2 x} = \sec^2 x,$

for $\qquad \dfrac{\sin \Delta x}{\Delta x} =$ unity, in the limit.

$\dfrac{d(\cot x)}{dx} = -\operatorname{cosec}^2 x.$ The proof is left to the student.

To prove that $\dfrac{d(\sec x)}{dx} = \sec x \tan x.$

With the same notation,

$$\Delta y = \sec(x + \Delta x) - \sec x$$

$$= \frac{\cos x - \cos(x + \Delta x)}{\cos x \cos(x + \Delta x)}$$

$$= \frac{2\sin\left(x + \dfrac{\Delta x}{2}\right)\sin\dfrac{\Delta x}{2}}{\cos x \cos(x + \Delta x)}.$$

$$\therefore \ \frac{\Delta y}{\Delta x} = \frac{\sin\left(x + \dfrac{\Delta x}{2}\right)}{\cos x \cos(x + \Delta x)} \times \frac{\sin\dfrac{\Delta x}{2}}{\dfrac{\Delta x}{2}}.$$

$\therefore$ proceeding to the limit, $\dfrac{dy}{dx} = \dfrac{\sin x}{\cos^2 x} = \sec x \cdot \tan x.$

The proof that $\dfrac{d(\operatorname{cosec} x)}{dx} = -\operatorname{cosec} x \cot x$ is left to the student.

EXAMPLE 1. *Find the differential coefficient of* $\tan^2 x.$

Let $\qquad y = \tan^2 x,$ and $u = \tan x,$

so that $\qquad\qquad\qquad y = u^2.$

$$\frac{dy}{du} = 2u, \quad \text{and} \quad \frac{du}{dx} = \sec^2 x.$$

$$\therefore \frac{dy}{dx} = \frac{dy}{du} \times \frac{du}{dx} = 2u \sec^2 x = 2 \tan x \cdot \sec^2 x.$$

Or, more shortly,

$$\frac{dy}{dx} = 2 \tan x \times \frac{d}{dx}(\tan x) \quad \left[\text{for } \frac{d}{dx}(p^2) = 2p\frac{dp}{dx} \right]$$

$$= 2 \tan x \cdot \sec^2 x.$$

EXAMPLE 2. *Differentiate* $\sin^2(5x - 3)$ *with respect to* $x.$

Let $\quad y = \sin^2(5x - 3),$ $u = \sin(5x - 3),$ and $v = 5x - 3,$

so that $\qquad\qquad y = u^2,$ and $u = \sin v;$

$$\frac{dy}{du} = 2u, \quad \frac{du}{dv} = \cos v, \quad \frac{dv}{dx} = 5.$$

$$\therefore \frac{dy}{dx} = \frac{dy}{du} \times \frac{du}{dv} \times \frac{dv}{dx} = 2u \cos v \times 5$$

$$= 10 \sin(5x - 3)\cos(5x - 3)$$

$$= 5 \sin(10x - 6).$$

Or, more shortly,

$$\frac{dy}{dx} = 2 \sin(5x - 3) \times \frac{d}{dx}[\sin(5x - 3)]$$

$$= 2 \sin(5x - 3) \times \cos(5x - 3) \times \frac{d}{dx}(5x - 3)$$

$$= 2 \sin(5x - 3) \times \cos(5x - 3) \times 5, \text{ as before.}$$

EXAMPLES III. b.

Find the differential coefficient of :

1. $\sin 2x$.   **2.** $\cos 3x$.   **3.** $\sin(2x-4)$.   **4.** $\cos(5-3x)$

5. $\sin\dfrac{x}{2}$.   **6.** $\cos\dfrac{x}{a}$.   **7.** $\tan\dfrac{x}{5}$.   **8.** $\sec\dfrac{x}{3}$.

9. $\tan 5x$.   **10.** $\tan(3x-5)$.   **11.** $\cot\left(\dfrac{\pi+x}{3}\right)$.

12. $\sec(3x-7)$.   **13.** $\operatorname{cosec}(9-4x)$.   **14.** $\sin^2 x$.

15. $\sin^2 5x$.   **16.** $\cos^2 x$.   **17.** $\sin^2(x+2)$.

18. $\cos^2(3x-4)$.   **19.** $\cos^3(4-3x)$.   **20.** $\tan^2(5x-6)$.

21. $\cos x^{\circ}$.   **22.** $\tan(x^{\circ}+45^{\circ})$.   **23.** $(1-\sin x)^2$.

24. $(1-\cos x)^3$.   **25.** $\dfrac{1}{1+\tan x}$.   **26.** $\sin^3 x$.   **27.** $\sin^3 2x$.

28. $\sin 8x\cos 2x$.   **29.** $\sin 2x\sin 4x$.   **30.** $\sqrt{\cos 3x}$.

31. $\sqrt[3]{\sin 3x}$.   **32.** $\sin^2 x+\cos^2 x$.   **33.** $\dfrac{1}{1+\sin x}$.

34. $\dfrac{1}{3-4\cos x}$.   **35.** $\dfrac{\cos x}{\cos x+\sin x}$.   **36.** $\dfrac{1}{\sec x-\tan x}$.

37. $\dfrac{1+\cos 2x}{1-\cos 2x}$.   **38.** Prove that $\dfrac{d}{dx}\left(\dfrac{\cos x+\sin x}{\cos x-\sin x}\right)=\sec^2\left(\dfrac{\pi}{4}+x\right)$.

39. $\sin x=\cos\left(\dfrac{\pi}{2}-x\right)$. Hence, assuming the value of $\dfrac{d}{dx}(\sin x)$, deduce the value of $\dfrac{d}{dx}(\cos x)$.

40. $\tan x=\dfrac{\sin x}{\cos x}$. Use the formula for $\dfrac{d}{dx}\left(\dfrac{u}{v}\right)$ to find the value of $\dfrac{d}{dx}(\tan x)$.

41. Deduce the value of $\dfrac{d}{dx}(\operatorname{cosec} x)$ from that of $\dfrac{d}{dx}(\sec x)$.

42. $\operatorname{cosec} x = \dfrac{1}{\sin x}$. Hence deduce the value of $\dfrac{d}{dx}(\operatorname{cosec} x)$ from that of $\dfrac{d}{dx}(\sin x)$.

43. Find the differential coefficient of $\dfrac{1 - \cos x}{1 + \cos x}$, and deduce that of $\dfrac{1 - \sin x}{1 + \sin x}$.

44. Differentiate $\sin mx \cdot \cos nx$ with respect to x, (1) by taking it as a product, (2) by taking it as equal to $\frac{1}{2}[\sin(m+n)x + \sin(m-n)x]$.

45. If $y = \sin x$, prove that

(a) $\dfrac{d^2 y}{dx^2} = \sin(\pi + x)$; (b) $\dfrac{d^3 y}{dx^3} = \sin\left(\dfrac{3\pi}{2} + x\right)$.

Deduce the value of $\dfrac{d^n y}{dx^n}$.

46. If $y = \cos x$, find the 2nd, 3rd, and 4th derivatives of y, and deduce the nth.

47. If $y = \sec x + \tan x$, prove that $\dfrac{dy}{dx} = y \sec x$.

EXAMPLES III. c.

[These may be taken orally.]

Write down, or read off, the value of:

1. $\dfrac{d}{dx}(u^2)$. 2. $\dfrac{d}{dx}(x+1)^6$. 3. $\dfrac{d}{dx}(3x+1)^5$.

4. $\dfrac{d}{dx}\left(\dfrac{1}{1+x}\right)$. 5. $\dfrac{d}{dx}\left(\dfrac{1}{1-x}\right)$. 6. $\dfrac{d}{dx}(1-x)^3$.

7. $\dfrac{d}{dx}(x-1)^{-3}$. 8. $\dfrac{d}{dx}\left(\dfrac{1}{1+x}\right)^2$. 9. $\dfrac{d}{dx}(\sin^2 x)$.

10. $\dfrac{d}{dx}(\sin 2x)$. 11. $\dfrac{d}{dx}(\tan kx)$. 12 $\dfrac{d}{dx}(\operatorname{cosec} mx)$.

13. $\dfrac{d}{dx}\left(\tan\dfrac{x}{a}\right)$. **14.** $\dfrac{d}{dx}\sin(2x+3)$. **15.** $\dfrac{d}{dx}(1-x)^{-4}$.

16 $\dfrac{d}{dx}(\sin^2 2x)$. **17.** $\dfrac{d}{dx}(ax^2+bx+c)^3$. **18.** $\dfrac{d}{dx}\left(\dfrac{1}{ax^2+bx+c}\right)$.

19. $\dfrac{d}{dx}(3-x)^{\frac{1}{3}}$ **20.** $\dfrac{d}{dx}(1+3x)^{\frac{1}{3}}$. **21.** $\dfrac{d}{dx}(1-x)^{-6}$.

22. $\dfrac{d}{dx}\left(\dfrac{1}{1-\cos x}\right)$. **23.** $\dfrac{d}{dx}\left(\dfrac{1}{1-\cos 2x}\right)$. **24.** $\dfrac{d}{dx}\left(\dfrac{1}{1+\tan x}\right)$.

25. $\dfrac{d}{dx}\left(\dfrac{1}{1+2x}\right)$. **26.** $\dfrac{d}{dx}\left(\dfrac{1}{5x-1}\right)$. **27.** $\dfrac{d}{dx}\left(\dfrac{1}{1-6x}\right)$.

APPROXIMATIONS

The side (a) of a square is increased by a small quantity, Δa; *to find the approximate increase in its area.*

The total increase in area $=(a+\Delta a)^2-a^2=2a\Delta a+(\Delta a)^2$.

But, Δa being a small quantity, $(\Delta a)^2$ is very small, and may be neglected;

$$\therefore \text{ the required increase}=2a\Delta a.$$

The student should illustrate this with a diagram, and notice the small *relative* magnitude of the area neglected.

To find the approximate error in using $\cos\theta$ *instead of* $\cos(\theta+\Delta\theta)$ *where* $\Delta\theta$ *is a small angle.*

The error $=\cos\theta-\cos(\theta+\Delta\theta)=2\sin\left(\theta+\dfrac{\Delta\theta}{2}\right)\sin\dfrac{\Delta\theta}{2}$

$$=\Delta\theta\sin\theta \text{ approx.}; \text{ for when } \dfrac{\Delta\theta}{2} \text{ is small,}$$

$\sin\dfrac{\Delta\theta}{2}=\dfrac{\Delta\theta}{2}$, and $\sin\left(\theta+\dfrac{\Delta\theta}{2}\right)=\sin\theta$ approx.

Or, $\cos(\theta+\Delta\theta)=\cos\theta\cos\Delta\theta-\sin\theta\sin\Delta\theta$

$$=\cos\theta-\Delta\theta\sin\theta; \text{ for } \cos\Delta\theta=1, \text{ and}$$
$$\sin\Delta\theta=\Delta\theta \text{ when } \Delta\theta \text{ is a small angle}$$

$\therefore$ the required error is $\Delta\theta\sin\theta$, as before.

We might obtain this by means of the calculus.

For if $\quad\quad\quad\quad y = \cos\theta, \dfrac{dy}{d\theta} = -\sin\theta$ ....(1)

But if Δy is the error in y,

$$\frac{\Delta y}{\Delta\theta} = \frac{dy}{d\theta} + \rho,$$

where ρ is a quantity which vanishes in the limit;

$$\therefore\ \Delta y = \Delta\theta\left(\frac{dy}{d\theta} + \rho\right)$$

$$= \Delta\theta\,.\,\frac{dy}{d\theta}\ \text{approx.}$$

$$= -\Delta\theta\,.\,\sin\theta, \text{ from (1).}$$

The error is negative, for as an angle increases, its cosine diminishes.

EXAMPLES III. d.

1. The side of a square is 4 ft. 1 in., and its area is taken to be 16 sq. ft. Find (1) the approximate error, (2) the percentage error.

2. Find the approximate error in using $\sin\theta$ instead of $\sin(\theta + a)$, where a is a small angle.

3. In a triangle, from the data A, B, b, the length of the side a is found; find the approximate error in this value if $A + \Delta A$ is the true value of the angle at A, ΔA being small, and the other data being correct.

4. Given in a triangle the sides a, b, and the included angle C, the third side (c) is calculated. Find the approximate error in this value of c, if $C + \Delta C$ is the true value of the angle at C, ΔC being small.

5. Find the approximate area between two concentric circles of radii $a + \Delta a$, and a, where Δa is small.

6. Given the sides a, b of a triangle, and the included angle C, its area is calculated. Find the approximate error in this area if $C + \Delta C$ is the true value of the angle at C, the values of a and b being correct.

7. If x be taken instead of the true value $x + \Delta x$, show that the percentage error is less than $\dfrac{100 \cdot \Delta x}{x}$.

8. Find the approximate percentage error in taking $9\cdot06$ as the value of $(3\cdot01)^2$.

9. Find the approximate change in the value of $5x^2 - 8x + 3$ as x changes from 2 to $2\cdot001$.

10. The space, s feet, described by a moving body in time t is given by the formula $s = 4t + 9t^2$; find the approximate space passed over during the $\frac{1}{20}$th of a second after 5 seconds of motion, and deduce the average velocity during that interval of time.

· CHAPTER IV.

A DIFFERENTIAL COEFFICIENT CONSIDERED AS A RATE-MEASURER.

If a particle describes space s in time t, to prove that its velocity at the end of time $t = \dfrac{ds}{dt}$.

Let Δt be a small increment of time (t), and let v and v' be the greatest and least velocities of the particle during this time Δt. If Δs is the space described in time Δt,

Δs cannot be greater than $v\,\Delta t$, nor less than $v'\Delta t$. $(s = vt)$

$\therefore \dfrac{\Delta s}{\Delta t}$ „ „ „ v, „ v'.

But as Δt continually diminishes, v and v' continually approach and ultimately become equal to the velocity at time t.

$\therefore$ proceeding to the limit, $\dfrac{ds}{dt} =$ the velocity at time t.

Thus we see that $\dfrac{ds}{dt}$ is the rate at which s is changing with respect to t; and of course, in the same way, $\dfrac{dy}{dx}$ is the rate at which y is changing with respect to x.

Geometrical Interpretation.

Let ABC be a portion of the graph of $y = f(x)$, and draw the tangent BT to the curve at the point B.

Then at B, $\dfrac{dy}{dx} = \tan \phi$, the slope of the tangent.

$\therefore$ the slope of the tangent to a curve at any point gives the rate at which the ordinate (y) at that point is increasing with respect to the abscissa (x).

This shows clearly that the *differential coefficient of a constant is zero*, for if y is constant, its rate of change is zero.

$$\therefore \frac{dy}{dx} = 0.$$

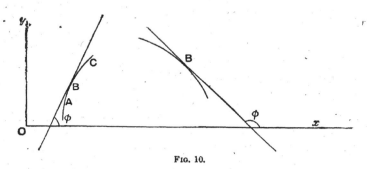

FIG. 10.

If v is the velocity of a particle at time t, to prove that its acceleration at time $t = \dfrac{dv}{dt} = \dfrac{d^2 s}{dt^2} = v \dfrac{dv}{ds}.$

If Δv is the increment of velocity in the increment of time Δt, and f, f' are the greatest and least values of the acceleration during that time,

Δv cannot be greater than $f \Delta t$, nor less than $f' \Delta t$. $(v = ft)$

$\therefore \dfrac{\Delta v}{\Delta t}$,,   ,, f,   ,,   . f'.

But as Δt continually diminishes, f and f' continually approach and ultimately become equal to the acceleration at time t.

$\therefore$ proceeding to the limit, $\dfrac{dv}{dt} = $ the acceleration at time t.

Also, $\dfrac{dv}{dt} = \dfrac{d}{dt}\left(\dfrac{ds}{dt}\right) = \dfrac{d^2 s}{dt^2}.$

Again, acceleration $= \dfrac{dv}{dt} = \dfrac{dv}{ds} \times \dfrac{ds}{dt} = v\dfrac{dv}{ds}.$

Second method of proving that acceleration $= \dfrac{dv}{dt} = \dfrac{d^2s}{dt^2}$.

By definition, acceleration = rate of change of velocity

$$\quad\quad\quad\text{''}\quad\quad\text{''}\quad\quad \frac{ds}{dt}\ \text{(or } v)$$

$$= \frac{d^2s}{dt^2}\ \left(\text{or } \frac{dv}{dt}\right).$$

EXAMPLE. *If* $y = \dfrac{x^3}{10}$, *find the rate at which y is changing with respect to x when* $x = 2$.

Geometrically. Plot the graph of $y = \dfrac{x^3}{10}$ near the point where $x = 2$, say for the values 0, 1, 2, 3, 4 of x.

Draw the tangent (as accurately as possible) at the point where $x = 2$.

The slope of this tangent is the rate required.

N.B.—The same unit must be used for x and y.

Analytically. $\dfrac{dy}{dx} = \dfrac{3x^2}{10} = \dfrac{3 \times 2^2}{10}$ when $x = 2$.

$$\therefore \text{ the required rate} = 1\cdot2.$$

NOTE. In the case of a straight line graph whose equation is $y = mx + c$,

$$\frac{dy}{dx} = m, \text{ the slope of the line.}$$

$\therefore$ the rate at which y is increasing with respect to x is given by the slope of the straight line, and is therefore constant.

EXAMPLE. *A hollow cone, whose semi-vertical angle is* 30°, *is held with its axis vertical and vertex downwards, and water is poured into it at the steady rate of 3 cubic feet per minute. Find the rate at which the depth (measured along the axis) of water is increasing when that depth is 3 feet.*

Let V be the volume of the water in the cone, when x is the depth, and r the radius of the surface of the water,

Then $V = \frac{1}{3}\pi r^2 x = \frac{\pi}{3} x^3 \tan^2 30° = \frac{\pi x^3}{9}$.

$$\therefore \frac{dV}{dt} = \frac{\pi x^2}{3} \cdot \frac{dx}{dt}.$$

But, by hypothesis, $\frac{dV}{dt} = 3$ c. ft. per min.

$$\therefore \frac{dx}{dt} = \frac{9}{\pi x^2},$$

i.e. the depth x is increasing at the rate of $\frac{9}{\pi x^2}$ ft. per minute.

$$\therefore \text{ the rate required} = \frac{1}{\pi} = \cdot 31831 \text{ ft. per min. approx.}$$

$$= 3 \cdot 82 \text{ inches per min. approx.}$$

EXAMPLES IV.

1. If the space-time equation of a particle moving in a straight line is $s = t^3$ (ft.-sec. units being used), find the velocity and acceleration of the particle after it has been moving for 5 secs.

2. If the space-time equation of a particle moving in a straight line is $s = a + bt^3$, prove that its acceleration varies as t.

3. The speed of a moving point changes from 34 ft. per sec. at the end of 3 secs. to 85 ft. per sec. at the end of 8 secs.; what is its average acceleration during this time?

4. The side of a square increases uniformly at the rate of 2 in. per sec.; what is the rate of increase of its area when the side is 5 in. long?

5. _A particle moves in a straight line, and its distance (in feet) from a fixed point at time t secs. is given by the equation $s = 4 + 10t - t^2$.

   Find (1) its velocity when it has been in motion for 2 seconds;

   (2) its distance from the fixed point when its velocity is zero;

   (3) its acceleration.

   Describe the motion of the particle.

6. A particle moves in a straight line, and its distance from a fixed point at time t is given by the equation $x = a \cos(\omega t)$.

   Find (1) its velocity at time t; (2) its maximum and minimum velocities.

   Prove that its acceleration varies as its distance from the fixed point, and is towards that point.

7. A particle moves in a straight line, and its distance from a fixed point at time t is given by the equation

   $$s = a + bt + ct^2 + t^3.$$

   Find its velocity at any time t, and prove that it is subject to a variable acceleration.

8. The velocity-space equation of a particle moving in a straight line is $v^2 = 16s$. Find its acceleration.

9. A circular metal plate is beaten out, always remaining circular in form, its radius increasing at the steady rate of $\frac{1}{10}$ in. per minute: find the rate of increase of its area with respect to the number of inches in its radius, (1) when its radius is x in.; (2) when its radius is 4 in.

10. A hollow vertical cylinder of radius 9 inches has water poured into it at the rate of 1 c. foot per minute. Find the rate at which the depth of water increases.

11. If the radius of a circle increases at a uniform-rate, prove that its area will increase at a rate which varies as its radius.

12. If the radius of a circle increases at the rate of one inch per second, find the rate at which its area is increasing when the radius is one foot.

13. If the area of a circle increases at the rate of 5 sq. ft. per sec., find the rate at which the radius increases, and find the value of this rate when the circumference is 10 feet.

14. If water is poured into an inverted hollow cone whose semi-vertical angle is 30°, so that its depth (measured along the axis) increases at the rate of 1 inch per second, find the rate at which the volume of water is increasing when the depth is 2 feet.

15. A reservoir, 60 ft. by 100 ft. at the top and 40 ft. by 80 ft. at the bottom, has plane sloping ends and sides. If it is 10 feet deep, find an expression for the area of the surface of the water when it is x feet deep.

If water is pumped into it so that its depth increases at the rate of one-tenth of an inch per hour, find the rate at which the area of the surface is increasing when the water is 3 feet deep.

16. A hollow pyramid, of height h, on a square base of side a, is inverted and held with its base horizontal. If water is poured into it at the rate of b cubic feet per second, prove that its depth (x) increases at the rate of $\dfrac{h^2 b}{a^2 x^2}$ feet per second.

17. A straight line AB cuts off with two fixed straight lines Ox, Oy at right angles a constant area of 100 sq. in. If OA increases at the rate of 2 in. per sec., find the rate at which OB is decreasing, (1) when OA = x, (2) when OA = 8 in.

18. A cylinder with a sliding piston contains 1000 cu. ft. of gas at a pressure of 10 lb. per sq. in. Find the rate of increase of the volume (V) with respect to the pressure (p) when the piston moves slowly.

19. Find the slope of the tangent at any point on the curve $s = 4t + 16t^2$, and also at the point where $t = 2$. If s is the distance, in feet, of a body from a fixed point after t secs., interpret the meaning of these slopes.

What would be the meaning of the slope of the chord joining any two points on the curve ?

20. Find the average increase of y per unit increase of x if $y = x^2 - 5x$, while x increases from 2·5 to 5.

21. If $y = 4x - 3x^2$, find the rate of change of y with respect to x when $x = \cdot5$.

CHAPTER V.

EXAMPLES ON MAXIMA AND MINIMA.

A right circular cylinder is inscribed in a sphere of given radius a : to find the maximum value of its curved surface.

If $2x$ is the altitude of the cylinder, the radius of its base

$$= \sqrt{a^2 - x^2}.$$

∴ its curved surface $= 2\pi \sqrt{a^2 - x^2} \cdot 2x \ldots\ldots(1) \quad (2\pi rh)$

$$= \text{S suppose.}$$

We have to find the maximum value of S.

In a case of this kind negative values of $\sqrt{a^2 - x^2}$ are inadmissible, hence we see that S is a maximum when S^2 is a maximum.

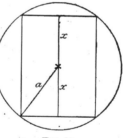

Fig. 11.

$$S^2 = 16\pi^2(a^2x^2 - x^4).$$

$$\therefore \frac{d(S^2)}{dx} = 16\pi^2(2a^2x - 4x^3) = 32\pi^2x(a^2 - 2x^2).$$

$$\therefore \frac{d(S^2)}{dx} = 0 \quad \text{when} \quad x = 0 \quad \text{or} \quad \frac{a}{\sqrt{2}}.$$

We need not trouble about $\frac{d^2(S^2)}{dx^2}$, for $x = 0$ evidently makes S a minimum.

$$\therefore \text{S is a maximum when } x = \frac{a}{\sqrt{2}}.$$

$$\therefore \text{ from (1), the maximum value of } S = 4\pi\sqrt{a^3 - \frac{a^2}{2}} \cdot \frac{a}{\sqrt{2}}$$

$$= 2\pi a^2.$$

In physical problems, such as the above, it is seldom necessary to use a second derivative. Common sense determines which value of the variable gives a maximum and which gives a minimum.

The above problem may be solved without the use of the Calculus.

Let $x = a\cos\theta,$ $(x < a)$

so that $S = 4\pi\sqrt{a^2 - a^2\cos^2\theta} \cdot a\cos\theta = 4\pi a^2\sin\theta\cos\theta$

$$= 2\pi a^2 \sin 2\theta.$$

$\therefore$ S is a maximum when $\sin 2\theta$ is greatest,

i.e. when $\sin 2\theta = 1$;

and the maximum value of S is $2\pi a^2$, as before.

EXAMPLES V.

[The cones and cylinders in the following examples may be taken as "right" and "circular."]

1. Find the maximum volume of a cone which can be inscribed in a sphere of radius a.

   [First find the volume of the cone in terms of its height, x.]

2. A cylinder is inscribed in a sphere of radius a: find its maximum volume.

   [Find the volume of the cylinder in terms of its height.]

3. Find the area of the greatest rectangle which can be inscribed in a circle of radius a.

4. The slant side of a cone is of given length l; find its maximum volume.

5. From the corners of a rectangular piece of cardboard, 16 inches by 6 inches, squares (side x) are cut out, and the edges are then turned up so as to form a box: find the value of x when its content is a maximum.

6. If the sides of a rectangle are 3 in. and 5 in. long, show that the greatest rectangle which can be drawn, so that its sides pass through the angular points of the given rectangle, is a square whose area is 32 sq. in.

[Let one side of the required rect. make an angle θ with one side of the given rect., and work in terms of θ.]

7. AB is a straight line cutting the axes of x and y at A and B, and passing through the fixed point (a, a). Find the minimum length of AB.

[Let the $\angle OAB = \theta$, and find the length of AB in terms of a and θ.]

8. Find the elevation (to the nearest half-degree) of the shortest ladder which would reach from the ground to the wall of a house over an obstacle 8 ft. high at a distance of 5 ft. from the wall.

9. Find the maximum value of $a \cos \theta + b \sin \theta$.

10. In a cone of height h and base-radius a, a cylinder is inscribed. Find the radius of the cylinder when its *total* surface is a maximum. From your result, prove that h must be greater than $2a$ if a maximum surface can be obtained.

11. A cylinder is inscribed in a cone of height h and base-radius a. Find its maximum volume.

12. If a rectangle is inscribed in a quadrant of a circle (radius a), *i.e.* with two of its sides along the bounding radii and one of its angular points on the arc, find its maximum area.

13. Through a given point (h, k) a straight line is drawn to meet the axes of co-ordinates in A and B. Prove that the minimum value of AB is $(h^{\frac{2}{3}} + k^{\frac{2}{3}})^{\frac{3}{2}}$.

14. A cistern of depth h, standing on the ground, is kept full of water, and a small hole is drilled in one side at a depth x feet below its top edge. If the water leaves this opening with a horizontal velocity of $\sqrt{2gx}$ ft. per sec., find the position of the hole when the water strikes the ground at a maximum distance from the cistern.

15. By post-office regulations the combined length and girth of a parcel must not exceed 6 feet. Find, to the nearest thousandth of a cubic foot, the maximum volume of a cylindrical parcel.

16. A man on one bank of a river, 3 miles broad, wishes to reach a point 6 miles down' the other bank. If he can travel in a motor-boat at 8 miles per hour, and bicycle at 10 miles per hour, find how far from his destination he must land on the opposite bank in order to reach it in a minimum time.

17. A straight line AB cuts the axes of x and y at A and B, and passes through the fixed point (a_1, a_2). Find the minimum value of the area of the triangle OAB.

18. On AB, one side of a rectangle ABCD, a semi-circle is described externally; given the perimeter of the figure, prove that when its area is a maximum; the radius of the semi-circle is equal to BC.

[Find the area of the figure in terms of the radius.]

19. In the portion of the parabola $y^2 = 16x$ bounded by the curve and the double ordinate whose equation is $x = 27$, a rectangle is inscribed having two sides parallel to the axis of x. Find its maximum area.

20. A cylindrical open bucket is made from 200 sq. inches of thin metal. Find its maximum volume to the nearest cubic inch.

CHAPTER VI.

MISCELLANEOUS EXAMPLES OF DIFFERENTIATION

To prove that $\dfrac{d(\sin^{-1}x)}{dx} = \dfrac{1}{\sqrt{1-x^2}}$.

Let $y = \sin^{-1}x$, so that $\sin y = x$.

$$\therefore \ \cos y = \frac{dx}{dy} \quad \left(\text{or we might say } \cos y \cdot \frac{dy}{dx} = 1\right)$$

$$\therefore \ \frac{dy}{dx} = \frac{1}{\dfrac{dx}{dy}} = \frac{1}{\cos y} = \frac{1}{\sqrt{1-x^2}}.$$

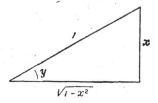

To prove that $\dfrac{d(\cos^{-1}x)}{dx} = \dfrac{-1}{\sqrt{1-x^2}}$.

Let $y = \cos^{-1}x$, so that $\cos y = x$.

$$\therefore \ -\sin y = \frac{dx}{dy} \quad \left(\text{or } -\sin y \cdot \frac{dy}{dx} = 1\right)$$

$$\therefore \ \frac{dy}{dx} = \frac{1}{\dfrac{dx}{dy}} = -\frac{1}{\sin y} = \frac{-1}{\sqrt{1-x^2}}.$$

N.B. $\dfrac{d(\sin^{-1}x)}{dx} + \dfrac{d(\cos^{-1}x)}{dx} = 0.$

We might see the truth of this from the fact that the angles $\sin^{-1}x$ and $\cos^{-1}x$ are complementary, so that

$$\sin^{-1}x + \cos^{-1}x = \frac{\pi}{2}.$$

$$\therefore \frac{d(\sin^{-1}x)}{dx} + \frac{d}{dx}(\cos^{-1}x) = 0, \text{ for } \frac{\pi}{2} \text{ is constant.}$$

To prove that $\dfrac{d\left(\tan^{-1}\mathbf{x}\right)}{d\mathbf{x}} = \dfrac{1}{1+\mathbf{x}^2}.$

Let $y = \tan^{-1}x$, so that $\tan y = x.$

$$\therefore \sec^2 y = \frac{dx}{dy} \qquad \therefore \frac{dy}{dx} = \frac{1}{\dfrac{dx}{dy}} = \cos^2 y$$

$$= \frac{1}{1+x^2}. \quad \text{(See the figure.)}$$

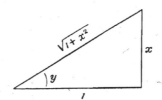

In the same way it can be proved that $\dfrac{d(\cot^{-1}\mathbf{x})}{d\mathbf{x}} = \dfrac{-1}{1+\mathbf{x}^2}$

To prove that $\dfrac{d(\sec^{-1}\mathbf{x})}{d\mathbf{x}} = \dfrac{1}{\mathbf{x}\sqrt{\mathbf{x}^2-1}}.$

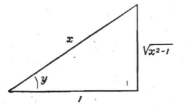

Let $y = \sec^{-1}x$, so that $\sec y = x$

$$\therefore \sec y \tan y = \frac{dx}{dy}.$$

$$\therefore \frac{dy}{dx} = \frac{1}{\dfrac{dx}{dy}} = \frac{1}{\sec y \tan y}$$

$$= \frac{1}{x\sqrt{x^2-1}}. \quad \text{(See the figure.)}$$

In the same way it can be proved that

$$\frac{d(\operatorname{cosec}^{-1}x)}{dx} = \frac{-1}{x\sqrt{x^2-1}}.$$

In the following examples it is assumed that the student is acquainted with the following series :

$$e^x = 1 + x + \frac{x^2}{\lfloor 2} + \frac{x^3}{\lfloor 3} + \dots \text{ ad inf.}$$

$$a^x = 1 + x \log_e a + \frac{(x \log_e a)^2}{\lfloor 2} + \frac{(x \log_e a)^3}{\lfloor 3} + \dots \text{ ad inf}$$

$$\log_e(1+x) = \quad x - \frac{x^2}{2} + \frac{x^3}{3} - \frac{x^4}{4} \dots \text{ ad inf.} \quad (x \text{ being} < 1)$$

$$\log_e(1-x) = -x - \frac{x^2}{2} - \frac{x^3}{3} - \frac{x^4}{4} \dots \text{ ad inf.} \qquad \text{,,}$$

The series form of e^x and a^x are valid for all values of x.
The series for $\log_e(1 \pm x)$ are only valid when x is numerically less than unity.

Unless something is said to the contrary, $\log x$ is always taken to be $\log_e x$ in the Calculus.

To prove that $\dfrac{d(e^x)}{dx} = e^x$.

Let $y = e^x$, and Δx, Δy be corresponding increments of x and y, so that $y + \Delta y = e^{x+\Delta x}$.

By subtraction, $\quad \Delta y = e^{x+\Delta x} - e^x$

$$= e^x [e^{\Delta x} - 1]$$

$$= e^x \left[\Delta x + \frac{(\Delta x)^2}{\lfloor 2} + \frac{(\Delta x)^3}{\lfloor 3} + \dots \right].$$

$$\therefore \frac{\Delta y}{\Delta x} = e^x \left[1 + \frac{\Delta x}{\lfloor 2} + \frac{(\Delta x)^2}{\lfloor 3} + \dots \right];$$

and proceeding to the limit, $\dfrac{dy}{dx} = e^x$.

This result might be obtained by writing down the series for e^x, and differentiating each term separately.

To prove that $\dfrac{d(e^{ax})}{dx} = ae^{ax}$.

The method of the preceding article may be used, or the following.

If $y = e^{ax}$, $\dfrac{dy}{dx} = e^{ax} \times \dfrac{d(ax)}{dx}$.   [Function of a function.]

$$= ae^{ax}.$$

To prove that $\dfrac{d(a^x)}{dx} = a^x \log_e a$.

Let $y = a^x$, and Δx, Δy be corresponding increments of x and y, so that $y + \Delta y = a^{x+\Delta x}$.

By subtraction, $\Delta y = a^{x+\Delta x} - a^x = a^x[a^{\Delta x} - 1]$

$$= a^x \left[\Delta x \log_e a + \frac{(\Delta x \log_e a)^2}{\underline{2}} + \frac{(\Delta x \log_e a)^3}{\underline{3}} + \dots \right].$$

$$\therefore \frac{\Delta y}{\Delta x} = a^x \left[\log_e a + \frac{\Delta x (\log_e a)^2}{\underline{2}} + \frac{(\Delta x)^2 (\log_e a)^3}{\underline{3}} + \dots \right];$$

and proceeding to the limit, $\dfrac{dy}{dx} = a^x \log_e a$.

This result might be obtained by differentiating the series for a^x.

To prove that $\dfrac{d(\log_e x)}{dx} = \dfrac{1}{x}$.

Let $y = \log_e x$, and Δx, Δy be corresponding increments of x and y, so that $y + \Delta y = \log_e(x + \Delta x)$.

By subtraction,

$$\Delta y = \log_e(x + \Delta x) - \log_e x$$

$$= \log_e \left(1 + \frac{\Delta x}{x} \right)$$

$$= \frac{\Delta x}{x} - \frac{1}{2} \left(\frac{\Delta x}{x} \right)^2 + \frac{1}{3} \left(\frac{\Delta x}{x} \right)^3 - \dots,$$

for $\dfrac{\Delta x}{x}$ will be less than unity if Δx is taken small enough.

$$\therefore \quad \frac{\Delta y}{\Delta x} = \frac{1}{x} - \frac{\Delta x}{2x^2} + \frac{(\Delta x)^2}{3x^3} - \cdots ,$$

and proceeding to the limit, $\dfrac{dy}{dx} = \dfrac{1}{x}$.

To prove that $\quad \dfrac{d(\log_a x)}{dx} = \dfrac{1}{x \log_e a} = \dfrac{\log_a e}{x}.$

$$\log_a x = \log_e x \times \log_a e, \quad \text{or} \quad = \frac{\log_e x}{\log_e a}.$$

$$\therefore \frac{d(\log_a x)}{dx} = \log_a e \, \frac{d}{dx}(\log_e x), \quad \text{or} \quad \frac{1}{\log_e a} \frac{d}{dx}(\log_e x)$$

$$\left[\text{for } \log_a e \text{ and } \frac{1}{\log_e a} \text{ are both constant} \right].$$

$$\therefore \frac{d(\log_a x)}{dx} = \frac{\log_a e}{x}, \quad \text{or} \quad \frac{1}{x \log_e a}.$$

Logarithmic Differentiation. When an expression is in the form of a product or quotient, differentiation is often simplified by the use of logarithms.

To find the differential coefficient of the expression $\dfrac{(x-1)^{\frac{3}{2}}}{(x-4)^{\frac{5}{2}}(x-2)^{\frac{1}{2}}}.$

If the given expression $= y$,

$$\log y = \frac{3}{2}\log(x-1) - \frac{5}{2}\log(x-4) - \frac{1}{2}\log(x-2).$$

$$\therefore \frac{1}{y}\frac{dy}{dx} = \frac{3}{2(x-1)} - \frac{5}{2(x-4)} - \frac{1}{2(x-2)};$$

$$\therefore \frac{dy}{dx} = \frac{(x-1)^{\frac{3}{2}}}{(x-4)^{\frac{5}{2}}(x-2)^{\frac{1}{2}}}\left[\frac{3}{2(x-1)} - \frac{5}{2(x-4)} - \frac{1}{2(x-2)} \right].$$

To find the value of $\dfrac{dy}{dx}$ *when* $y = \dfrac{\sin^7 x}{\cos^5 x}$.

$$\log y = 7 \log (\sin x) - 5 \log (\cos x).$$

$$\therefore \ \frac{1}{y} \frac{dy}{dx} = \frac{7}{\sin x} \frac{d(\sin x)}{dx} - \frac{5}{\cos x} \frac{d(\cos x)}{dx}$$

$$= 7 \cot x + 5 \tan x;$$

$$\therefore \ \frac{dy}{dx} = \frac{\sin^7 x}{\cos^5 x} [7 \cot x + 5 \tan x].$$

EXAMPLES VL.

Prove the following :

1. $\dfrac{d\left(\sin^{-1}\dfrac{x}{a}\right)}{dx} = \dfrac{1}{\sqrt{a^2 - x^2}}.$

2. $\dfrac{d\left(\cos^{-1}\dfrac{x}{a}\right)}{dx} = \dfrac{-1}{\sqrt{a^2 - x^2}}.$

3. $\dfrac{d\left(\tan^{-1}\dfrac{x}{a}\right)}{dx} = \dfrac{a}{a^2 + x^2}.$

4. $\dfrac{d\left(\cot^{-1}\dfrac{x}{a}\right)}{dx} = \dfrac{-a}{a^2 + x^2}.$

5. $\dfrac{d\left(\sec^{-1}\dfrac{x}{a}\right)}{dx} = \dfrac{a}{x\sqrt{x^2 - a^2}}.$

6. $\dfrac{d\left(\operatorname{cosec}^{-1}\dfrac{x}{a}\right)}{dx} = \dfrac{-a}{x\sqrt{x^2 - a^2}}.$

Find $\dfrac{dy}{dx}$ in the following :

7. $y = \sin^{-1} 2x.$ 8. $y = \tan^{-1} 2x.$ 9. $y = \cos^{-1}\dfrac{x}{2}.$

10. $y = \sin^{-1}\sqrt{1 - x^2}.$ 11. $y = \sin^{-1}(1 - x).$

12. $y = \tan^{-1}\left(\dfrac{3x + 2}{4}\right).$ 13. $y = \cos^{-1}\left(\dfrac{a - x}{a}\right).$

14. $y = \dfrac{1 - \tan x}{\sec x}.$ 15. $y = \tan^{-1}\dfrac{x}{\sqrt{1 - x^2}}.$

16. $y = \sin^{-1}\left(\dfrac{1 - x^2}{1 + x^2}\right).$ 17. $y = \tan^{-1}\dfrac{1}{\sqrt{x^2 - 1}}.$

18. $y = \tan^{-1}\left(\dfrac{a+x}{1-ax}\right).$ **19.** $y = \log\left(\dfrac{x+a}{x-a}\right).$

20. $y = \log(\cos x).$ **21.** $y = x\log x.$ **22.** $y = \log\dfrac{x^2 - a^2}{x^2 + a^2}$

23. $y = \sin^{-1}\dfrac{x}{\sqrt{1+x^2}}.$ **24.** $y = \tan(\log x).$ **25.** $y = e^{ax}\sin bx.$

26. $y = e^x(1 - x^4).$ **27.** $y = e^x + e^{-x}.$

28. $y = \sec^{-1}\dfrac{a}{\sqrt{a^2 - x^2}}.$ **29.** $y = \sec^{-1}\dfrac{1}{2x^2 - 1}.$

Find the differential coefficient of each of the following expressions :

30. $\log(x + \sqrt{a^2 + x^2}).$ **31.** $\log(e^x + e^{-x}).$ **32.** $e^{\frac{x}{a}} + e^{-\frac{x}{a}}$

33 $\dfrac{1 - \cos\dfrac{x}{2}}{1 + \cos\dfrac{x}{2}}$ **34.** $\dfrac{\sin x - \cos x}{\sin x + \cos x}.$ **35.** $\dfrac{\sin x}{1 - \cos x}.$

36. $\sin mx \sin nx.$ **37.** $\sin^3(ax^2 + b).$ **38.** $\sin(\log x).$

39. $y = \dfrac{\sqrt{ax(x - 3a)}}{\sqrt{x - 4a}}.$ **40.** $\tan^{-1}\dfrac{2x}{1 - x^2}.$

41. $x^x.$ (Take logs.)   **42.** $\left(\dfrac{x}{n}\right)^{nx}.$ **43.** e^{e^x}

44. $\log(x + \sqrt{1 + x^2}).$ **45.** $x\cos^{-1}x - \sqrt{1 - x^2}.$

46. $\tan^{-1}\dfrac{4x}{4 - x^2}.$ **47.** $\tfrac{1}{2}\cos^{-1}(2x^2 - 1).$

48. $(1 + x^2)\tan^{-1}x - x.$ **49.** $\log\dfrac{1 - \cos x}{1 + \cos x}.$

50. $\log(\sqrt{x + 1} + \sqrt{x - 1}).$ **51.** $\log\dfrac{1 - \sin x}{1 + \sin x}.$

52. $\dfrac{a^2}{2}\sin^{-1}\dfrac{x}{a} + \dfrac{x}{2}\sqrt{a^2 - x^2}.$ **53.** $\dfrac{x}{2}\sqrt{a^2 + x^2} + \dfrac{a^2}{2}\log(x + \sqrt{a^2 + x^2})$

54. $(a \sin^2 x + b \cos^2 x)^6$.   **55.** $\dfrac{\sin^2 x}{\cos^2 x}$.   (Take logs.)

56. $\dfrac{(x-1)^{\frac{2}{3}}}{(x+1)^{\frac{1}{3}}(x-2)^{\frac{2}{3}}}$.   (Take logs.)   **57.** $\log(\log x)$.

58. $\tan^{-1}\sqrt{\dfrac{1-\cos x}{1+\cos x}}$.

59 If $y = \sqrt{\dfrac{1+x}{1-x}}$, prove that $\dfrac{dy}{dx} = \dfrac{y}{1-x^2}$.

60. Assuming that $\sin\theta = \theta - \dfrac{\theta^3}{\underline{3}} + \dfrac{\theta^5}{\underline{5}} - \dfrac{\theta^7}{\underline{7}} \dots$ *ad inf.*, deduce a series for $\cos\theta$.

61. Prove that $\dfrac{d^n \sin\theta}{d\theta^n} = \sin\left(\dfrac{n\pi}{2} + \theta\right)$.

62. If $y = \sin ax$, find the value of $\dfrac{d^n y}{dx^n}$.

63. Prove that $\dfrac{d^n \cos\theta}{d\theta^n} = \cos\left(\dfrac{n\pi}{2} + \theta\right)$.

64. If $y = \cos ax$, find the value of $\dfrac{d^n y}{dx^n}$.

65. If $y = a^x$, find the value of $\dfrac{d^n y}{dx^n}$.

66. If $y = e^{ax}$, find the value of $\dfrac{d^n y}{dx^n}$.

CHAPTER VII.

INTEGRATION CONSIDERED AS THE REVERSE OF DIFFERENTIATION.

THE student will learn, later on, that **Integration** means summation, but at first he will do well to look upon it as merely **the reverse process to that of differentiation.**

EXAMPLE. $\frac{d}{dx}(x^2) = 2x$; $\therefore$ the integral of $2x$ is x^2, or, as it is usually written, $\int 2x\,dx = x^2$.

The symbol $\int$ is a form of the old-fashioned **s.**

$\int 2x\,dx$ is read thus : the integral of $2x$ with respect to x.

EXAMPLES. $\frac{d}{dx}\left(\frac{x^n}{n}\right) = x^{n-1}$; $\therefore$ $\int x^{n-1}\,dx = \frac{x^n}{n}$.

$\frac{d}{dx}(\sin x) = \cos x$; $\therefore$ $\int \cos x\,dx = \sin x$.

$\frac{d}{dx}(\log x) = \frac{1}{x}$; $\therefore$ $\int \frac{dx}{x} = \log x$.

The differential coefficients of $ax^n + c$ and ax^n are the same, viz. nax^{n-1}.

The differential coefficients of $f(x) + c$ and $f(x)$ are the same, viz. $f'(x)$.

Hence, in integrating, we may always add an arbitrary constant.

Thus $\int x^3 dx = \dfrac{x^4}{4} + c$, where c may be any constant.

$\dfrac{x^4}{4}$ is called the *general* integral, or the *indefinite* integral of x^3

The expression to be integrated is called the integrand.

N.B.—**An integration can always be checked by differentiation.**

$$\int (ax^2 + bx + c)dx = \frac{ax^3}{3} + \frac{bx^2}{2} + cx.$$

Check. $\dfrac{d}{dx}\left(\dfrac{ax^3}{3} + \dfrac{bx^2}{2} + cx\right) = \dfrac{3ax^2}{3} + \dfrac{2bx}{2} + c = ax^2 + bx + c.$

The following are important, and should be committed to memory.

$\int x^n dx = \dfrac{x^{n+1}}{n+1}$ for all values of n except -1.

$\int x^{-1}dx = \int \dfrac{dx}{x} = \log_e x.$ $\qquad \int dx = x.$ $\qquad \int \dfrac{dx}{x^2} = -\dfrac{1}{x}.$

$\int \sin x \, dx = -\cos x.$ $\qquad \int \cos x \, dx = \sin x.$

$\int \sec^2 x \, dx = \tan x.$ $\qquad \int \operatorname{cosec}^2 x \, dx = -\cot x.$

$\int \sec x \tan x \, dx = \int \dfrac{\sin x \, dx}{\cos^2 x} = \sec x.$

$\int \operatorname{cosec} x \cot x \, dx = \int \dfrac{\cos x \, dx}{\sin^2 x} = -\operatorname{cosec} x.$ $\qquad \int e^x dx = e^x$

$\int a^x dx = \dfrac{a^x}{\log_e a}$, or $a^x \log_e e.$ $\qquad \int \dfrac{dx}{\sqrt{1-x^2}} = \sin^{-1}x$ or $-\cos^{-1}x.$

$\int \dfrac{dx}{\sqrt{a^2-x^2}} = \sin^{-1}\dfrac{x}{a}$ or $-\cos^{-1}\dfrac{x}{a}.$ $\quad \int \dfrac{dx}{1+x^2} = \tan^{-1}x$ or $-\cot^{-1}x.$

$\int \dfrac{dx}{a^2+x^2} = \dfrac{1}{a}\tan^{-1}\dfrac{x}{a}$ or $-\dfrac{1}{a}\cot^{-1}\dfrac{x}{a}.$

$$\int \frac{dx}{x\sqrt{x^2 - 1}} = \sec^{-1}x \text{ or } -\csc^{-1}x.$$

$$\int \frac{dx}{x\sqrt{x^2 - a^2}} = \frac{1}{a}\sec^{-1}\frac{x}{a} \text{ or } -\frac{1}{a}\csc^{-1}\frac{x}{a}.$$

Typical Examples. $\int (x-3)^4 dx = \frac{(x-3)^5}{5}.$

$$\int (3-x)^4 dx = -\frac{(3-x)^5}{5}$$

$$\int \sqrt{x+4}\, dx = \frac{2}{3}(x+4)^{\frac{3}{2}}. \qquad \int \sin 2\theta\, d\theta = -\frac{1}{2}\cos 2\theta.$$

$$\int \sin^2 \theta\, d\theta = \frac{1}{2}\int (1 - \cos 2\theta)\, d\theta = \frac{1}{2}\left(\theta - \frac{1}{2}\sin 2\theta\right).$$

$$\int \frac{x-4}{x-3} dx = \int \left(1 - \frac{1}{x-3}\right) dx = x - \log(x-3).$$

$$\int (2x-1)^2 = \frac{1}{6}(2x-1)^3. \qquad \int \frac{dx}{3x-7} = \frac{1}{3}\log(3x-7).$$

$$\int (ax+b)^n dx = \frac{(ax+b)^{n+1}}{a(n+1)}, \text{ except when } n = -1.$$

In that case, $\int (ax+b)^{-1}dx = \dfrac{\log(ax+b)}{a}.$

The student should carefully examine and check all the above.

EXAMPLES VII.

Integrate the following expressions with respect to x.

1. $3x^2$.   2. $2x$.   3. a.   4. $6x^2$.

5. $12x^3$.   6. $-6x^5$.   7. $14x^6$.   8. $3x^2 - 2x$.

9. ax^2.   10. $7bx^6$.   11. $4cx^4$.   12. $\dfrac{x}{2}$.

13. $\dfrac{x^2}{5}$.   14. $x^3 + x^2$.   15. $x^2 - ax$.   16. $x+1$.

17. $1-x$.   18. $2x-1$.   19. $4-2x$.   20. x^3-1.

21. $x^2 - a^2$.   22. x^{-4}.   23. $\dfrac{1}{x^2}$.   24. $\dfrac{1}{x^5}$.

25. x^{-11}.   26. $-\dfrac{1}{x^6}$.   27. x^{-1}.   28. $ax^2 + bx + c$.

29. $-5x^{-4}$.   30. $x(x-1)$.   31. $\dfrac{x+a}{2}$.   32. $x^2 + 1 + \dfrac{1}{x^2}$.

33. $\dfrac{2}{5}x^{\frac{4}{3}}$.   34. $x^{\frac{4}{3}}$.   35. $\sqrt{x}$.   36. $\dfrac{1}{\sqrt{x}}$.

37. $\dfrac{x+1}{x}$.   38. $\dfrac{x^2+1}{x^2}$.   39. $\dfrac{x^3+1}{x}$.   40. $\dfrac{1}{x^{\frac{1}{3}}}$.

41. $\dfrac{ax^2 + bx + c}{x}$.   42. $(x+2)^3$.   43. $(x-1)^4$.

44. $\dfrac{1}{x-3}$.   45. $(x+3)^5$.   46. $\dfrac{1}{(x-1)^2}$.

47. $\dfrac{1}{(x+2)^7}$.   48. $(x+7)^5$.   49. $(x-1)^8$.

50. $(ax+b)^4$.   51. $(2x-3)^5$.   52. $(1-x)^7$.

53. $\dfrac{1}{(2x-3)^2}$.   54. $(1-2x)^5$.   55. $\dfrac{1}{1+x}$.

56. $\sqrt{x-3}$.   57. $\dfrac{1}{(ax+b)^3}$.   58. $\dfrac{1}{ax+b}$.

59. $\dfrac{1}{(x+2)^2}$.   60. $\sqrt{ax+b}$.   61. $\dfrac{1}{2x+3}$.

62. $\dfrac{1}{\sqrt{x-4}}$.   63. $\dfrac{1}{(3-x)^2}$.   64. $\dfrac{1}{3-x}$.

65. $\dfrac{1}{a-bx}$.   66. $(1-x)^3$.   67. $(1-2x)^7$.

68. $(3-x)^4$.   69. $\dfrac{1}{(3-2x)^2}$.   70. $(3-4x)^3$.

71. $\dfrac{x^2 + bx + c}{x^2}$.   72. $\dfrac{x^2 + x - 2}{x^4}$.   73. $(x+1)(x-2)$.

74. $(x+1)(x-2)^2$. 75. $x(x-1)^3$. 76. $\dfrac{x+1}{x-1}$.

77. $\dfrac{3x+1}{x-2}$. 78. $\dfrac{x^2-x-3}{x-2}$. 79. $\dfrac{5}{x^2}$.

80. $\dfrac{1}{(1+x)^2}$. 81. $\dfrac{1}{(1-x)^2}$. 82. $\sin\dfrac{x}{2}$.

83. $\sin\left(\dfrac{\pi}{2}+x\right)$. 84. $\sec^2\dfrac{x}{2}$. 85. $\sec\dfrac{x}{2}\tan\dfrac{x}{2}$.

86. $\cos\left(\dfrac{\pi}{2}-x\right)$. 87. $\sec^2 2x$. 88. $\sin 2x$.

89. $\cos\left(\dfrac{\pi}{2}-2x\right)$. 90. $\sin(\pi-2x)$. 91. $\sin ax$.

92. $\cos bx$. 93. $\dfrac{\sin ax}{\cos^2 ax}$. 94. $\sin(x-a)$.

95. $\cos(a-x)$. 96. $\sec^2(a-x)$. 97. $\dfrac{1}{\sin^2(a+bx)}$.

98. $2\sin 4x\cos 2x$. 99. $\sin 3x\sin x$. 100. $\sin mx\cos nx$.

101. $\sin^2 x$. 102. $\cos^2 x$. 103. $\cos^2 ax$.

104. $\sin^2 ax$. 105. $\sin^2(ax+b)$. 106. $\cos^2(bx+c)$.

107. e^{-x}. 108. $e^{\frac{x}{a}}$. 109. e^x+e^{-x}.

110. $e^{\frac{x}{a}}+e^{-\frac{x}{a}}$ 111. 10^x. 112. e^{-2x}.

113. $\dfrac{e^{2x}+1}{e^x}$. 114. $(e^x+e^{-x})^2$

CHAPTER VIII.

DIFFERENTIALS. · CHANGING THE VARIABLE.
USE OF PARTIAL FRACTIONS.

Differentials. The student has learnt that $\frac{dy}{dx}$ is a single expression in which dy and dx cannot be separated, just as we cannot separate sin and θ in $\sin \theta$.

With the usual notation, $\frac{dy}{dx} =$ the limiting value of $\frac{\Delta y}{\Delta x}$, Δy and Δx being finite quantities.

i.e. $\frac{\Delta y}{\Delta x} = \frac{dy}{dx} + \rho$, where ρ is a quantity which vanishes in the limit.

$$\therefore \Delta y = \left(\frac{dy}{dx} + \rho \right) \Delta x.$$

$\therefore \Delta y$ continually approaches $\frac{dy}{dx} \cdot \Delta x$ as we decrease Δx; and when Δx is indefinitely small we may say that

$$\Delta y = \frac{dy}{dx} \Delta x.$$

In such a case Δy and Δx are called **differentials**.

This equation is sometimes written $Dy = \frac{dy}{dx} \cdot Dx$, it being understood that Dx and Dy are indefinitely small.

Thus, if $\qquad y = f(x), \qquad Dy = f'(x) \cdot Dx.$

$\qquad\qquad\qquad y = 2ax^2, \qquad Dy = 4ax \cdot Dx.$

$\qquad\qquad\qquad y = \sin \theta, \qquad Dy = \cos \theta \cdot D\theta.$

A more common notation, which we shall adopt, is

$$dy = \frac{dy}{dx} \times dx,$$

it being remembered that dy and dx when used separately must be regarded as finite quantities, but just as small as we please.

EXAMPLES.   If $y = x^3$, $\qquad dy = \frac{dy}{dx} \times dx = 3x^2 . dx.$

$\qquad y = \sin 2x, \qquad dy = \quad , \quad = 2 \cos 2x . dx.$

$\qquad y = ax^2 + bx, \quad dy = \quad , \quad = (2ax + b) dx.$

Changing the variable. By changing the variable the integral can often be reduced to one of the standard forms (see p. 64).

EXAMPLE 1.   *To find the value of* $\int x^2 (ax^3 + b)^5 dx.$

Let $ax^3 + b = u$, so that $3ax^2 dx = du$.

Then, by substitution,

$$\int x^2 (ax^3 + b)^5 dx = \int x^2 . u^5 . \frac{du}{3ax^2} = \frac{1}{3a} \int u^5 du$$

$$= \frac{u^6}{18a} = \frac{(ax^3 + b)^6}{18a}.$$

EXAMPLE 2.   *To integrate* $\sin \theta \cos^7 \theta$ *with respect to* θ.

Let $\cos \theta = u$, so that $-\sin \theta . d\theta = du$.

Then, by substitution,

$$\int \sin \theta \cos^7 \theta \, d\theta = -\int \sin \theta . u^7 \frac{du}{\sin \theta}$$

$$= -\int u^7 \, du = -\frac{u^8}{8} = -\frac{\cos^8 \theta}{8}.$$

Important.   In Example 1 above,

$x^2 \times$ a constant is the differential coefficient of $ax^3 + b$.

This is why we take $ax^3 + b = u$, the new variable.

Similarly, in Example 2, $\sin \theta \times (-1)$ is the differential coefficient of $\cos \theta$, and hence we take $\cos \theta = u$, the new variable.

In general terms, if the integral is of the form $\int f'(x)\left[f(x)\right]^n dx$, where $f'(x)$ is the differential coefficient of $f(x)$, then take $f(x) = u$, the new variable.

EXAMPLE 3. $\int \sin^3 x\,dx = \int (1 - \cos^2 x)\sin x\,dx.$

Let $\cos x = u$, so that $-\sin x\,dx = du$, .etc.

Expressions can often be rationalised by a suitable change of variable.

EXAMPLE 4. $\qquad y = \int \dfrac{x^2 dx}{\sqrt{3x + 4}}.$

Let $3x + 4 = u^2$, so that $3dx = 2u\,du$, and $x = \dfrac{u^2 - 4}{3}.$

By substitution, we have

$$y = \frac{2}{27}\int (u^2 - 4)^2 du = \frac{2}{27}\int (u^4 - 8u^2 + 16)\,du, \text{ etc.}$$

To find the value of $\int \sqrt{a^2 - x^2}\,dx.$

Let $x = a \sin\theta$, so that $dx = a \cos\theta\,d\theta.$

By substitution,

$$\int \sqrt{a^2 - x^2}\,dx = \int a \cos\theta\,.\,a \cos\theta\,d\theta = a^2\int \cos^2\theta\,d\theta,$$

$$= \frac{a^2}{2}\int 1 + \cos 2\theta)\,d\theta = \frac{a^2}{2}\left(\theta + \frac{\sin 2\theta}{2}\right)$$

$$= \frac{a^2}{2}(\theta + \sin\theta \cos\theta)$$

$$= \frac{a^2}{2}\left[\sin^{-1}\left(\frac{x}{a}\right) + \frac{x}{a}\sqrt{1 - \frac{x^2}{a^2}}\right]$$

$$= \frac{a^2}{2}\sin^{-1}\left(\frac{x}{a}\right) + \frac{x}{2}\sqrt{a^2 - x^2}.$$

To prove that $\displaystyle\int \frac{dx}{\sqrt{x^2 + a^2}} = \log(x + \sqrt{x^2 + a^2})$.

Let $u - x = \sqrt{x^2 + a^2}$, so that $u^2 - 2ux = a^2$.

Differentiating, we have $2u\,du - 2x\,du - 2u\,dx = 0$;

$$\therefore (u - x)\,du = u\,dx.$$

$\therefore$ by substitution, $\displaystyle\int \frac{dx}{\sqrt{x^2 + a^2}} = \int \frac{(u - x)\,du}{(u - x)u} = \int \frac{du}{u}$

$$= \log u = \log(x + \sqrt{x^2 + a^2}).$$

Integrate the function $x^3\sqrt{2x - 3}$.

Let $2x - 3 = u^2$, so that $2dx = 2u\,du$ and $x = \dfrac{u^2 + 3}{2}$.

Then, by substitution,

$$\int x^3\sqrt{2x - 3}\,dx = \int \frac{(u^2 + 3)^3}{8}u^2\,du$$

$$= \frac{1}{8}\int (u^8 + 9u^6 + 27u^4 + 27u^2)\,du$$

$$= \frac{1}{8}\left(\frac{u^9}{9} + \frac{9u^7}{7} + \frac{27u^5}{5} + 9u^3\right)$$

$$= \frac{u^3}{8}\left(\frac{u^6}{9} + \frac{9u^4}{7} + \frac{27u^2}{5} + 9\right)$$

$$= \frac{(2x - 3)^{\frac{3}{2}}}{8}\left[\frac{(2x - 3)^3}{9} + \frac{9}{7}(2x - 3)^2 + \frac{27}{5}(2x - 3) + 9\right]$$

$$= \text{etc.}$$

INTEGRATION BY MEANS OF PARTIAL FRACTIONS.

To find the value of $\displaystyle\int \frac{dx}{x^2 - a^2}$.

Here we must use partial fractions.

Let $\dfrac{1}{x^2 - a^2} \equiv \dfrac{A}{x - a} + \dfrac{B}{x + a}$, so that $1 \equiv A(x + a) + B(x - a)$.

Putting $x = a$, we have $A = \dfrac{1}{2a}$.

„ $x = -a$, „ $B = -\dfrac{1}{2a}$.

$$\therefore \frac{1}{x^2 - a^2} = \frac{1}{2a}\left[\frac{1}{x - a} - \frac{1}{x + a}\right].$$

$$\therefore \int \frac{dx}{x^2 - a^2} = \frac{1}{2a}\int \frac{dx}{x - a} - \frac{1}{2a}\int \frac{dx}{x + a}$$

$$= \frac{1}{2a}\left[\log(x - a) - \log(x + a)\right]$$

$$= \frac{1}{2a}\log\frac{x - a}{x + a}.$$

It is assumed here that $x > a$.

If $x < a$, $x - a$ is negative. In this case we must take the given expression as equal to

$$\frac{1}{2a}\left[-\frac{1}{a - x} - \frac{1}{a + x}\right].$$

The integral is then

$$\frac{1}{2a}\left[\log(a - x) - \log(a + x)\right]$$

$$= \frac{1}{2a}\log\left(\frac{a - x}{a + x}\right).$$

To find the value of $\displaystyle\int \frac{dx}{6x^2 + 5x - 4}$.

$$\frac{1}{6x^2 + 5x - 4} \equiv \frac{1}{(3x + 4)(2x - 1)} \equiv \frac{A}{3x + 4} + \frac{B}{2x - 1} \text{ suppose}$$

$$\therefore 1 \equiv A(2x - 1) + B(3x + 4).$$

Putting $x = \dfrac{1}{2}$, we have $B = \dfrac{2}{11}$.

„ $x = -\dfrac{4}{3}$, „ $A = -\dfrac{3}{11}$.

$$\therefore \int \frac{dx}{6x^2 + 5x - 4} = \frac{2}{11}\int \frac{dx}{2x - 1} - \frac{3}{11}\int \frac{dx}{3x + 4}$$

$$= \frac{1}{11}\log(2x - 1) - \frac{1}{11}\log(3x + 4)$$

$$= \frac{1}{11}\log\frac{2x - 1}{3x + 4}.$$

EXAMPLES VIII.

Using differentials, write down the value of dy in the following cases :

1. $y = 5x - 3.$ 2. $y = x^3.$ 3. $y = 2x^2 + 4.$

4. $y^2 = x.$ 5. $y = \sin 2x.$ 6. $y = \sin^2 x.$

7. $y = \tan^2 x.$ 8. $y = \sec^2 \frac{x}{2}.$ 9. $y = \sqrt{1 + x}.$

10. If $u^2 - 2ux = a^2$, prove that $dx = \frac{u - x}{u}\cdot du.$

11. If $u - x = \sqrt{x^2 - a^2}$, find the value of $dx.$

Integrate the following expressions :

12. $(3x - 4)^7.$ 13. $2x(x^2 - 3)^4.$ 14. $\frac{2ax + b}{ax^2 + bx + c}.$

15. $(2x + 3)^7.$ 16. $(3 - 4x)^3.$ 17. $(1 - 2x)^{\frac{1}{2}}.$

18. $x(x^2 + 1)^3.$ 19. $x^2(x^3 - 1)^2.$ 20. $x^{n-1}(x^n + a^n)^3.$

21. $\frac{1}{\sqrt{1 - 2x}}.$ 22. $\frac{x}{1 - x^2}.$ 23. $\frac{x}{1 + 3x^2}.$ 24. $\frac{1}{4x - 3}.$

25. $\tan x.$ 26. $\cot x.$ 27. $\sin 2x.$ 28. $\sin\frac{x}{2}.$

29. $\sec^2 6x.$ 30. $\sec^2\frac{x}{3}.$ 31. $\csc^2 3x.$ 32. $\frac{x}{\sqrt{1 - x^2}}.$

33. $\frac{2ax + b}{\sqrt{ax^2 + bx + c}}.$ 34. $\frac{2x - 5}{\sqrt{x^2 - 5x + 3}}.$ 35. $\frac{2x + 1}{\sqrt{x^2 + x - 2}}.$

36. $\dfrac{x}{1+x^2}$.   **37.** $x\sqrt{ax^2+b}$.   **38.** $(x-2)(x^2-4x-3)^{\frac{1}{2}}$

39. $\dfrac{\sin x}{3+5\cos x}$.   **40.** $\dfrac{x}{(x^2-1)^{\frac{3}{2}}}$.   **41.** $\cos^3 x$.   **42.** $\cos^3 2x$.

43. $(x+a)(x^2+2ax+b)^{\frac{3}{2}}$.   **44.** $\dfrac{x^3}{ax^4+b}$.

45. $\tan 2x$.   **46.** $x^3\sqrt{x^4-1}$.

47. $\sqrt{1-x^2}$.   (See p. 70.)   **48.** $\dfrac{1}{\sqrt{1+x^2}}$.   (See p. 71.)

49. $\dfrac{1}{\sqrt{x^2-1}}$.   **50.** $x^2\sqrt{3x-1}$.   **51.** $\sin^2\dfrac{x}{2}$.

52. $\dfrac{\cos x}{\sqrt{\sin x}}$.   **53.** $\sec^2 x\,e^{\tan x}$.   **54.** $\dfrac{1}{x^2+4}$.   **55** $\dfrac{1}{1+9x^2}$.

56. $\dfrac{2(1-x)}{1+2x}$.   **57.** $\dfrac{x+1}{x^2+2x+2}$.   **58.** $x^4(x^5+1)^7$.   **59.** $\dfrac{x}{\sqrt{x^2+1}}$.

60. $(x^2+1)^{-\frac{1}{2}}$. $\left(\text{Take } u=\dfrac{1}{x^2}\right)$.   **61.** $(x^2+a^2)^{-\frac{1}{2}}$.   **62.** $\dfrac{1}{\sqrt{x^2+4}}$.

63. $\sqrt{9-x^2}$.   **64.** $\dfrac{1}{\sqrt{x^2-4}}$.   **65.** $\dfrac{\cos x}{4+\sin^2 x}$.   **66.** $\dfrac{\sec^2 x}{4+\tan^2 x}$.

67. $\dfrac{x}{\sqrt{5x+4}}$.   (Let $5x+4=u^2$.)   **68.** $\dfrac{x^2}{\sqrt{x+1}}$.   **69.** $\dfrac{x}{\sqrt{x+2}}$.

Find the value of

70. $\displaystyle\int\dfrac{dx}{1-x^2}$.   **71.** $\displaystyle\int\dfrac{x\,dx}{x-1}$.   **72.** $\displaystyle\int\dfrac{dx}{x^2-5x+6}$.

73. $\displaystyle\int\dfrac{dx}{x^2-1}$.   **74.** $\displaystyle\int\dfrac{x\,dx}{1-4x^2}$.   **75.** $\displaystyle\int\dfrac{dx}{(x-1)(x+3)}$.

76. $\displaystyle\int\dfrac{5\,dx}{(3x-1)(2x+1)}$.   **77.** $\displaystyle\int\dfrac{dx}{(x-a)(x-b)}$.

78. $\int \dfrac{(3x^2 - 12x + 11)}{(x-1)(x-2)(x-3)} dx.$ **79.** $\int \dfrac{x^2 + 2x - 5}{(x^2-1)(x-2)} dx.$

80. $\int \dfrac{2dx}{x(x^2-1)}.$ **81.** $\int \dfrac{dx}{x^2(x^2+1)}.$

82. $\int \dfrac{dx}{x^2(4x^2+1)}.$ **83.** $\int \dfrac{5x+1}{x^2+4} dx.$ **84.** $\int \dfrac{x^2-5}{2\sqrt{x}} dx.$

85. $\int \sqrt{2a+x} \, dx.$ **86.** $\int \dfrac{dx}{x^2 - 3x + 2}.$ **87.** $\int \dfrac{x^3 + a^3}{2x} dx.$

88. $\int \dfrac{dx}{x^2+3}.$ **89.** $\int \dfrac{dx}{x\sqrt{x^2-4}}.$ **90.** $\int \dfrac{3x+1}{x^2+1} dx.$

91. $\int \dfrac{1+2x}{\sqrt{1-x^2}} dx.$ **92.** $\int \dfrac{1+x}{2+x^2} dx.$ **93.** $\int \tan^2 2x \, dx.$

94. $\int \sin^3 \dfrac{x}{2} \, dx.$ **95.** $\int \dfrac{\cos x}{(1-\sin x)^2} dx.$

CHAPTER IX.

DEFINITE INTEGRALS.

LET $\int y\,dx = f(x)$. Then we may define the definite integral $\int_b^a y\,dx$ as being equal to $f(a) - f(b)$.

It is often written thus : $\int_b^a y\,dx = \left[f(x) \right]_b^a = f(a) - f(b)$.

a and b are called the *limits*, a being the *upper limit*, b the *lower limit*.

EXAMPLES.

$$\int_1^3 x^3 dx = \left[\frac{x^4}{4} \right]_1^3 = \frac{1}{4}[3^4 - 1^4] = 20.$$

$$\int_1^5 \frac{dx}{x} = \left[\log_e x \right]_1^5 = \log_e 5 - \log_e 1 = \log_e 5 = \frac{\log_{10} 5}{\log_{10} e} = \frac{\cdot 6990}{\cdot 4343}, \text{ etc.}$$

$$\int_0^1 (1 - x)^4 dx = -\frac{1}{5} \left[(1 - x)^5 \right]_0^1 = \frac{1}{5}.$$

In using definite integrals, no arbitrary constant need be introduced, as we see from the following :

If $\quad \int y\,dx = f(x) + c,$

$$\int_b^a y\,dx = \left[f(x) + c \right]_b^a = f(a) + c - f(b) - c = f(a) - f(b).$$

$$\int_b^a y\,dx = \int_0^a y\,dx - \int_0^b y\,dx.$$

For if $\int y\,dx = \phi(x)$,

$$\int_0^a y\,dx - \int_0^b y\,dx = \phi(a) - \phi(0) - [\phi(b) - \phi(0)]$$
$$= \phi(a) - \phi(b)$$
$$= \int_b^a y\,dx.$$

Also note that $\int_a^b f(x)\,dx = \phi(b) - \phi(a) = \cdots [\phi(a) - \phi(b)]$

$$= -\int_b^a f(x)\,dx.$$

Changing the variable.

To find the value of $\int_0^a \sqrt{a^2 - x^2}\,dx$.

Let $x = a \sin\theta$, so that $dx = a \cos\theta\,.\,d\theta$.

When $x = a$, $\sin\theta = 1$, and $\theta = \dfrac{\pi}{2}$.

„ $x = 0$, $\sin\theta = 0$, „ $\theta = 0$.

$\therefore$ by substitution, $\int_0^a \sqrt{a^2 - x^2}\,dx = \int_0^{\frac{\pi}{2}} a\cos\theta\,.\,a\cos\theta\,d\theta$

$$= a^2 \int_0^{\frac{\pi}{2}} \cos^2\theta\,d\theta = \frac{a^2}{2} \int_0^{\frac{\pi}{2}} (1 + \cos 2\theta)\,d\theta$$

$$= \frac{a^2}{2}\left[\theta + \frac{\sin 2\theta}{2}\right]_0^{\frac{\pi}{2}} = \frac{a^2}{2}\cdot\frac{\pi}{2} = \frac{\pi a^2}{4}.$$

N.B.—We must change the limits when we change the variable.

EXAMPLES IX.

Find the value of each of the following:

1. $\int_1^4 x\,dx$.

2. $\int_0^2 x^7\,dx$.

3. $\int_1^3 (x - 2)\,dx$.

4. $\int_0^3 (3 - x)\,dx$.

5. $\int_4^5 \dfrac{dx}{x - 3}$ (2 decimal places).

6. $\displaystyle\int_0^{\frac{\pi}{2}} \sin 3\theta\, d\theta.$ **7.** $\displaystyle\int_5^9 \frac{dx}{\sqrt{x-5}}.$ **8.** $\displaystyle\int_{-3}^6 \sqrt{x+3}\, dx.$

9. $\displaystyle\int_0^4 \frac{dx}{\sqrt{6x+1}}.$ **10.** $\displaystyle\int_{-\frac{\pi}{4}}^{\frac{\pi}{4}} \frac{d\theta}{\cos^2\theta}.$ **11.** $\displaystyle\int_0^{\frac{\pi}{2}} \sin^2\theta\, d\theta.$

12. $\displaystyle\int_0^a x\sqrt{a^2 - x^2}\, dx.$ **13.** $\displaystyle\int_1^2 x(x^2-1)^5 dx.$

14. $\displaystyle\int_0^{\frac{\pi}{2}} \sin 3\theta \cdot \cos\theta \cdot d\theta.$ **15.** $\displaystyle\int_{-\frac{\pi}{2}}^{\frac{\pi}{2}} \sec^2\frac{\theta}{2}\, d\theta.$

16. $\displaystyle\int_0^3 x\sqrt{x^2+16}\, dx.$ **17.** $\displaystyle\int_0^4 \frac{x\, dx}{\sqrt{x^2+9}}.$ **18.** $\displaystyle\int_0^{\frac{\pi}{2}} \cos^2\theta\, d\theta$

19. $\displaystyle\int_0^2 \sqrt{4-x^2}\, dx.$ **20.** $\displaystyle\int_0^{\frac{\pi}{2}} \sin^3\theta \cos\theta\, d\theta.$

21. $\displaystyle\int_0^{\frac{\pi}{2}} \sin^3 x\, dx.$ **22.** $\displaystyle\int_{-\frac{\pi}{2}}^{\frac{\pi}{2}} \frac{\cos x}{\sqrt{\sin x}}\, dx.$ **23.** $\displaystyle\int_0^2 \frac{dx}{x^2+4}.$

24. $\displaystyle\int_{-\frac{\pi}{2}}^{\frac{\pi}{2}} \sin^2\frac{x}{2}\, dx.$ **25.** $\displaystyle\int_0^6 \frac{x-1}{x+1}\, dx.$

26. $\cos^4\theta = \dfrac{1}{4}(2\cos^2\theta)^2 = \dfrac{1}{4}(1+\cos 2\theta)^2$

$$= \frac{1}{4}(1 + 2\cos 2\theta + \cos^2 2\theta).$$ Use this to prove that

$$\int_{-\frac{\pi}{2}}^{\frac{\pi}{2}} \cos^4\theta \cdot d\theta = \frac{3\pi}{8}.$$

27. $\sin^5\theta = \sin\theta(1 - \cos^2\theta)^2.$ Use this, taking $\cos\theta = u$, to

prove that $\displaystyle\int_0^{\frac{\pi}{2}} \sin^5\theta \cdot d\theta = \frac{8}{15}.$

28. Transform the integral $\int_0^a x^2(a^2 - x^2)^{\frac{1}{2}}dx$ by the substitution $x = a\cos\theta$, and find its value.

29. Prove that $\int_0^\infty \dfrac{x^4 dx}{(1 + x^2)^5} = \dfrac{1}{16}\int_0^{\frac{\pi}{2}} \sin^4 2u\, du$

$$= \dfrac{1}{64}\int_0^{\frac{\pi}{2}}(1 - \cos 4u)^2 du = \dfrac{3\pi}{256}.$$

(Let $x = \tan u$.)

30. Prove that $\int_0^\infty \dfrac{x^5 dx}{(1 + x^2)^6} = \dfrac{1}{60}.$

CHAPTER X.

EASY DIFFERENTIAL EQUATIONS.

THE student should study the following examples on integration carefully. He will learn later on that if

$$f(y) \cdot dy = \phi(x) \cdot dx, \quad \text{then} \int f(y) \, dy = \int \phi(x) \, dx + k,$$

where k is *any* constant.

The reverse process of differentiation will establish this property, and we will apply it to a few simple cases. In each example k is *any* constant.

EXAMPLE 1. *If* $y \dfrac{dy}{dx} = ax$, *or* $y \, dy = ax \, dx$,

then by integration, $\dfrac{y^2}{2} = \dfrac{ax^2}{2} + k.$

For $\dfrac{d}{dx}\left(\dfrac{y^2}{2}\right) = \dfrac{2y}{2}\dfrac{dy}{dx} = y\dfrac{dy}{dx}$, and $\dfrac{d}{dx}\left(\dfrac{ax^2}{2} + k\right) = \dfrac{2ax}{2} = ax.$

Now $\dfrac{y^2}{2} = \int y \, dy$, and $\dfrac{ax^2}{2} + k = \int ax \, dx$, where k is any constant.

∴ we have shown that if $y \, dy = ax \, dx$,

then $\int y \, dy = \int ax \, dx + k.$

EXAMPLE 2.   *If* $(y^2 + 2y)\dfrac{dy}{dx} = 4x^3$.   *or* $(y^2 + 2y)dy = 4x^3 dx$,

$\quad$ *then by integration* $\dfrac{y^3}{3} + y^2 = x^4 + k$.

For $\dfrac{d}{dx}\left(\dfrac{y^3}{3} + y^2\right) = (y^2 + 2y)\dfrac{dy}{dx}$,   and $\dfrac{d}{dx}(x^4 + k) = 4x^3$.

EXAMPLES X. a.

1. If $\dfrac{dy}{dx} = y$, prove that $\log y = x + a$, where a is any constant.

2. If $\dfrac{dx}{dt} = \sqrt{x}$, prove that $4x = (t + a)^2$, where a is any constant.

3. Given that $\dfrac{dv}{ds} = \dfrac{s^2}{v}$, prove that $3v^2 = 2s^3$, if $v = 0$ when $s = 0$.

4. Given that $\dfrac{dy}{dx} = y + 3$, prove that $x + k = \log_e(y + 3)$, k being any constant.

5. If $x\dfrac{dy}{dx} = (1 - y)^2$, prove that $y = 1 - \dfrac{1}{\log_e kx}$, where k is any constant.

6. If $(1 + x)\dfrac{dy}{dx} = 1 + y$, prove that $\log_e\left(\dfrac{1 + y}{1 + x}\right)$ is constant.

7. If $\dfrac{dy}{dx} = \sqrt{1 - y^2}$, prove that $y = \sin(x + k)$, k being any constant.

8. If $x\dfrac{dy}{dx} = x - y$, prove that $xy = \dfrac{x^2}{2} + k$, k being any constant.
$\left[\text{Use the formula } u\dfrac{dv}{dt} + v\dfrac{du}{dt} = \dfrac{d}{dt}(uv).\right]$

MISCELLANEOUS APPLICATIONS OF THE CALCULUS.

Given that the velocity of a point moving in a straight line varies as t the time, investigate its motion when s = 0 and t = 0 simultaneously.

By hypothesis, $\dfrac{ds}{dt} = at$, where a is constant.

$\therefore$ integrating, we have $s = a \displaystyle\int t\,dt = \dfrac{at^2}{2} + c$, where c is constant.

But when $t = 0$, $s = 0$; $\therefore c = 0$.

$$\therefore s = \dfrac{at^2}{2}.$$

Again, the acceleration of the point $= \dfrac{dv}{dt} = \dfrac{d(at)}{dt} = a.$

(See p. 45.)

$\therefore$ the point is subject to a constant acceleration a in the direction of motion.

A point has a uniform acceleration f in a straight line. Given that when s = 0, t = 0, and v = u, find the relation between s and t.

By hypothesis, $\dfrac{dv}{dt} = f$; $\therefore$ integrating, $v = ft + c$, where c is constant.

But $v = u$ when $t = 0$; $\therefore c = u$; $\therefore \mathbf{v = u + ft}$.

Again, $\dfrac{ds}{dt} = v = u + ft$; $\therefore$ integrating, $s = ut + \frac{1}{2}ft^2 + k$, where k is constant.

But when $s = 0$, $t = 0$; $\therefore k = 0$;

$$\therefore \mathbf{s = ut + \tfrac{1}{2}ft^2}.$$

To find the relation between v and s in the above case.

First Method. $\dfrac{dv}{ds} = \dfrac{\frac{dv}{dt}}{\frac{ds}{dt}} = \dfrac{f}{v}$; $\therefore v\,dv = f\,ds$.

$\therefore$ integrating, $\dfrac{v^2}{2} = fs + c$, where c is constant.

But when $s = 0$, $v = u$; $\therefore \dfrac{v^2}{2} = fs + \dfrac{u^2}{2}.$

$$\therefore \mathbf{v}^2 = \mathbf{u}^2 + 2\mathbf{fs}.$$

Second Method. $\dfrac{d^2s}{dt^2} = f;$ $\therefore \dfrac{ds}{dt} \cdot \dfrac{d^2s}{dt^2} = f \dfrac{ds}{dt}.$

$\therefore$ integrating, $\dfrac{1}{2} \cdot \overline{\dfrac{ds}{dt}}\Big|^2 = fs + c$, as before.

EXAMPLES X. b.

1. If a body moves in a straight line and changes its distance from a fixed point in that line at the rate of $2t$ ft. per sec., t denoting the time in secs.; find the distance of the body from the fixed point after 5 secs., given that the body passes through the fixed point when $t = 2$.

2. The velocity of a body in feet per sec. is given by the equation $v = (2t + 3)^2$. Find the space travelled in the 3rd second.

3. If the acceleration of a point is 4 ft.-sec. units in a constant direction, use the Calculus to find its velocity (v) after 6 secs., and the space (s) described in that time, given that when $t = 0$, $s = 0$, and $v = 3$ ft. per sec.

4. A point has an acceleration of 8 ft.-sec. units in a constant direction. Use the Calculus to find its velocity when it has described 21 feet, given that the point starts from rest.

5. If $at + b$ is the acceleration of a point moving in a straight line, find its velocity (v) after time t, and the space (s) described, given that when $t = 0$, $s = c$, and $v = u$.

6. A point is moving in a straight line such that the rate of increase of its distance (s) from a fixed point varies as the time (t). Express this as a differential equation, and find the value of s at time t, given that $s = a$ when $t = 0$, and the velocity of the point is equal to u when $t = 1$.

7. If the component velocities of a point parallel to the axes of co-ordinates are respectively $u \cos a$ and $u \sin a - gt$ at time t, find the co-ordinates of the point at time t, having given that when $t = 0$, $x = y = 0$.

8   $\dfrac{d^2x}{dt^2} = 0,\ \dfrac{d^2y}{dt^2} = f$ (f being constant) give the component accelerations at time t of a point parallel to the axes of co-ordinates. Given that when $t = 0$, $x = y = 0$, and $\dfrac{dx}{dt} = \dfrac{dy}{dt} = f$, find the co-ordinates of the point at time t.

9. Water is poured into a conical vessel held with its axis vertical, and it is found that the rate at which the depth (x) increases varies inversely as the depth at time t. Prove that the volume at time t varies as $t^{\frac{3}{2}}$.

10. Water flows through a hole in the bottom of a tank, and the rate at which the depth (x) of water decreases varies as the square root of the depth at time t. Given that $x = a$ when $t = 0$, and that when $x = 1$ the rate of decrease of depth $= 1$, find the depth of water at time t.

11. If the acceleration of a point moving in a straight line increases uniformly by 4 ft.-sec. units per sec., find its velocity after 5 secs., if the initial acceleration is 3 ft.-sec. units and the initial velocity is zero.

CHAPTER XI.

INTEGRATION BY PARTS.

IF u and v are functions of x,

$$\frac{d(uv)}{dx} = v\frac{du}{dx} + u\frac{dv}{dx}.$$

$\therefore$ integrating, we have

$$uv = \int\left(v\frac{du}{dx}\right)dx + \int\left(u\frac{dv}{dx}\right)dx;$$

i.e. $\int\left(u\frac{dv}{dx}\right)dx = uv - \int\left(v\frac{du}{dx}\right)dx.$

This is not an easy formula to remember. The following may help.

The integral of $p \cdot q = p \times$ the integral of q

$$-\int(\text{the diff. coefft. of } p) \times (\text{the integral of } q).$$

Perhaps the best plan for a beginner is to obtain the formula from the differential coefficient of a product each time he works an example.

EXAMPLE 1. $\int x\cos x\,dx = x\sin x - \int \sin x\,dx$

$$= x\sin x + \cos x.$$

Here $u = x, \quad \dfrac{dv}{dx} = \cos x,$

so that $\dfrac{du}{dx} = 1 \quad \text{and} \quad v = \sin x.$

EXAMPLE 2. $\int \sqrt{x^2+a^2}\,dx = x\sqrt{x^2+a^2} - \int\left[x\frac{d}{dx}(x^2+a^2)^{\frac{1}{2}}\right]dx$

$$\left(\text{Here } \frac{dv}{dx}=1, \text{ so that } v=x.\right)$$

$$= x\sqrt{x^2+a^2} - \int\frac{x^2}{\sqrt{x^2+a^2}}\,dx$$

$$= x\sqrt{x^2+a^2} - \int\frac{x^2+a^2-a^2}{\sqrt{x^2+a^2}}\,dx$$

$$= x\sqrt{x^2+a^2} - \int\sqrt{x^2+a^2}\,dx + a^2\int\frac{dx}{\sqrt{x^2+a^2}}.$$

$$\therefore\ 2\int\sqrt{x^2+a^2}\,dx = x\sqrt{x^2+a^2} + a^2\int\frac{dx}{\sqrt{x^2+a^2}}$$

$$= x\sqrt{x^2+a^2} + a^2\log(x+\sqrt{x^2+a^2}). \quad \text{(See p. 71.)}$$

$$\therefore\ \int\sqrt{x^2+a^2}\,dx = \frac{x}{2}\sqrt{x^2+a^2} + \frac{a^2}{2}\log(x+\sqrt{x^2+a^2}).$$

EXAMPLES XI.

Find the value of

1. $\int \log x\,dx.$ $\left(\text{Take } \log x=u, \text{ and } 1=\dfrac{dv}{dx}.\right)$

2. $\int xe^x\,dx.$ 3. $\int x^2e^x\,dx.$ 4. $\int x\sin ax\,dx.$

5. $\int x\cos ax\,dx.$ 6. $\int x^2\sin x\,dx.$ 7. $\int\sqrt{x^2-a^2}\,dx.$

8. $\int\sqrt{x^2+1}\,dx.$ 9. $\int\sqrt{x^2+25}\,dx.$ 10. $\int\log_{10}x\,dx.$

11. $\int\log(x+\sqrt{a^2+x^2})\,dx.$ 12. $\int x^2\sin 2x\,dx$

13. $\int \sin x \cos x \, dx$, (1) by parts ; (2) by taking it as $\dfrac{1}{2} \int \sin 2x \, dx$.
Account for any difference in the two results.

14. $\int xe^{ax} \, dx$.　　**15.** $\int \sqrt{a^2 - x^2} \, dx$.　　**16.** $\int x^5 \log x \, dx$.

17. $\int x \tan^{-1} x \, dx$.　　　　**18.** $\int \tan^{-1} x \, dx$.

CHAPTER XII.

AREAS AND VOLUMES.

Lᴇᴛ AB be part of a curve, and divide the abscissa (OC) of the point B into a very large number (n) of equal parts, OE, EG, GK, ..., each equal to Δx. Draw the ordinates EF, GH, ...

Complete the rectangles AOE, FEG, HGK, ...

Also   „   „   FEO, HGE, ... as shown in the diagram.

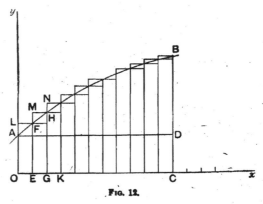

Fɪɢ. 12.

Let $OA = y_0$, $EF = y_1$, $GH = y_2$, ... $CB = y_n$.

The sum of the areas of the smaller-rectangles

$$= \Delta x(y_0 + y_1 + y_2 + ... + y_{n-1}).$$

The sum of the areas of the greater rectangles

$$= \Delta x(y_1 + y_2 + ... + y_n).$$

∴ the difference between the areas of the greater and smaller rectangles $= \Delta x(y_n - y_0) = \Delta x \cdot BD$.

Now let the number (n) of equal parts of OC be *increased indefinitely* (OC remaining fixed), so that Δx is *decreased indefinitely*.

Then $\Delta x \cdot BD$, the difference between the areas of the greater and smaller rectangles diminishes indefinitely, and ultimately vanishes, for Δx becomes zero.

∴ in the limit, the sum of the smaller rectangles

=   „   „   greater   „

= the area between the curve AB, the ordinate BC, the axes of co-ordinates.

To find the area contained between a curve, one of its ordinates, and the axes of co-ordinates.

Let $y = f'(x)$ be the equation of the curve AB (see Fig. 12), where $f'(x)$ is the differential coefficient of $f(x)$. Let $OC = a$, so that the ordinate $BC = f'(a)$.

Divide OC into a very large number (n) of equal parts, each equal to Δx, and draw ordinates at the points of division.

Then $OA = f'(0)$, $EF = f'(\Delta x)$, $GH = f'(2\Delta x) \ldots CB = f'(n\Delta x)$.

The area required (AOCB) = the limiting value of the areas of the strips AE, FG, HK, ... when n is indefinitely increased

= the limiting value of

$$\Delta x f'(0) + \Delta x f'(\Delta x) + \Delta x f'(2\Delta x) + \ldots + \Delta x f'(\overline{n-1}\,\Delta x)$$

= the limiting value of

$$\Delta x [f'(0) + f'(\Delta x) + f'(2\Delta x) + \ldots + f'(\overline{n-1}\,\Delta x)]. \ldots\ldots\ldots(1)$$

Now we see, from the definition of a differential coefficient,

that $$f'(x) = \frac{f(x + \Delta x) - f(x)}{\Delta x} + \rho,$$

where ρ is a quantity which vanishes in the limit.

$$\therefore \quad f'(0) = \frac{f(\Delta x) - f(0)}{\Delta x} + \rho_1,$$

$$f'(\Delta x) = \frac{f(2\Delta x) - f(\Delta x)}{\Delta x} + \rho_2,$$

$$f'(2\Delta x) = \frac{f(3\Delta x) - f(2\Delta x)}{\Delta x} + \rho_3,$$

$$\cdots\cdots\cdots\cdots\cdots\cdots\cdots\cdots\cdots\cdots\cdots$$

$$f'(\overline{n-1}\Delta x) = \frac{f(n\Delta x) - f(\overline{n-1}\Delta x)}{\Delta x} + \rho_n,$$

where ρ_1, ρ_2, ... are quantities which each vanish in the limit.

$\therefore$ by addition, $f'(0) + f'(\Delta x) + f'(2\Delta x) + ... + f'(\overline{n-1}\Delta x)$

$$= \frac{f(n\Delta x) - f(0)}{\Delta x} + \rho_1 + \rho_2 + ...$$

$\therefore \Delta x[f'(0) + f'(\Delta x) + f'(2\Delta x) + ... + f'(\overline{n-1}\Delta x)]$
$$= f(n\Delta x) - f(0) + \Delta x(\rho_1 + \rho_2 + ...).$$

$\therefore$ from (1), the area required (AOCB)

= the limiting value of $f(n\Delta x) - f(0) + \Delta x(\rho_1 + \rho_2 + ...)$

$= f(n\Delta x) - f(0)$ for Δx vanishes in the limit

$= f(a) - f(0)$ (for $n\Delta x = OC = a$)

$= \int_0^a f'(x)dx$, by the definition of a definite integral.

Important. The above shows that the process of integration is the equivalent of summation.

In this case the definite integral $\int_0^a f'(x)dx$ is equal to the limiting value of the *sum* of the strips AE, FG, etc.

The student will find it very useful to remember that $\int y\,dx =$ the sum of all small things like $y\,dx$.

From the preceding we see that, if $y = f(x)$ is the equation of the curve AB, and B is the point $(h,\,k)$,

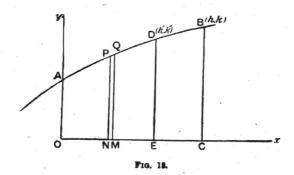

FIG. 18.

the area AOCB = the limiting value of the sum of the strips like PNMQ from $x = h$ to $x = 0$

$$= \int_0^h y\,dx; \text{ for the area of the elemental strip}$$
PNMQ $= y\Delta x$, approx., if $(x,\ y)$ are the co-ors. of P.

Also the area DECB = area AOCB – area AOED

$$= \int_0^h y\,dx - \int_0^{h'} y\,dx$$

$$= \int_{h'}^h y\,dx, \text{ if } (h',\ k') \text{ are the co-ors. of D.}$$

NOTE. If we take PQ as a straight line, PNMQ is a trapezium whose area is

$$\tfrac{1}{2}(PN + QM)NM,$$

which is equal to

$$\tfrac{1}{2}(2y + \Delta y)\Delta x$$

$$= y\Delta x, \text{ neglecting the product } \frac{\Delta x \cdot \Delta y}{2}$$
of two very small quantities

To find the area bounded by the curve $x^2 = 4ay$, the axis of x, and the ordinate at the point (h, k) on the curve.

The curve is a parabola, as shown in the diagram, and we have to find the area of OPAB.

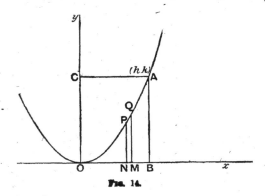

Fig. 14.

Take two points P, Q on the curve, near to one another, and let (x, y) be the co-ors. of P, and $(x + \Delta x, y + \Delta y)$ those of Q Draw the ordinates PN, QM.

The area of the elemental strip PNMQ $= y\Delta x$ approx.

If A denotes the area required, we may look upon PNMQ as the increment of area corresponding to the increment Δx.

$$\therefore \quad \Delta A = y\Delta x, \; i.e. \; \frac{\Delta A}{\Delta x} = y, \; \text{and in the limit} \; \frac{dA}{dx} = y \, ;$$

$\therefore$ by integration, $A = \int y\, dx$ between the proper limiting values of x.

These values of x are h and 0.

$$\therefore \; A = \int_0^h y\, dx = \frac{1}{4a}\int_0^h x^2 dx \quad (\text{for } x^2 = 4ay)$$

$$= \frac{1}{4a}\left[\frac{x^3}{3}\right]_0^h = \frac{h^3}{12a} = \frac{4ahk}{12a} \quad (\text{for } h^2 = 4ak)$$

$$= \frac{1}{3}hk = \frac{1}{3} \text{ of the rect. OBAC.}$$

To find the area of the circle $x^2 + y^2 = a^2$.

If (x, y) are the co-ors. of the point P, the area of the elemental strip $PNMQ = y\Delta x$ approx.

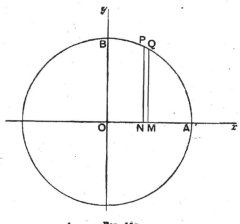

FIG. 15.

∴ the area of the quadrant OAB = the sum of all such strips from $x = a$ to $x = 0$.

∴ the area of the circle $= 4\int_0^a y\, dx = 4\int_0^a \sqrt{a^2 - x^2}\, dx$ *

$$= 4\frac{\pi a^2}{4} = \pi a^2.$$

To find the area of a circle, using polar co-ordinates.

Let (a, θ) be the co-ors. of the point P, and $(a, \theta + \Delta\theta)$ those of Q, Q being very near to P, so that the $\angle POQ = \Delta\theta$.

OPQ is approximately a $\triangle$.

$$\therefore \text{ its area} = \frac{a^2}{2}\sin\Delta\theta = \frac{a^2}{2}\Delta\theta,$$

for $\Delta\theta$ is a very small angle.

* For the working of this integral, see p. 70.

∴ the area of the quadrant OAB = the sum of all such sectors from $\theta = 0$ to $\theta = \dfrac{\pi}{2}$.

∴ the area of the circle $= 4 \displaystyle\int_0^{\frac{\pi}{2}} \tfrac{1}{2} a^2 d\theta = 2a^2 \Big[\theta \Big]_0^{\frac{\pi}{2}} = \pi a^2.$

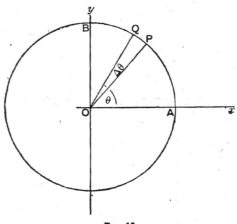

Fɪɢ. 16.

If part of an area falls in the 3ʳᵈ or 4ᵗʰ quadrant, the area of that part must be found separately, for in these quadrants y is negative and the calculated area will be negative. Before integrating, a graph (at least a rough one) of the curve should be drawn to see whether any part of the curve falls in the 3ʳᵈ or 4ᵗʰ quadrant.

Take as an example the curve $x^2 - 8x - y + 12 = 0.$

It is a parabola, for the term of the second degree is a perfect square.

When $y = 0$, $x^2 - 8x + 12 = 0$,   ∴ $x = 2$ or 6.

And when · $x = 0$, $y = 12.$

The equation may be written : $(x - 4)^2 = y + 4.$

∴ its vertex is at the point $(4, -4)$, and the curve lies as shown in the diagram.

The area of the part PAQ $= \int_2^6 (-y)\,dx$

(for y is negative throughout this part)

$= \int_2^6 (8x - x^2 - 12)\,dx = \left[4x^2 - \dfrac{x^3}{3} - 12x \right]_2^6$

$= 4(6^2 = 2^2) - \dfrac{1}{3}(6^3 - 2^3) - 12(6 - 2)$

$= 10\tfrac{2}{3}.$

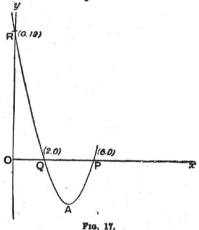

FIG. 17.

The area of the part ROQ $= \int_0^2 y\,dx = \int_0^2 (x^2 - 8x + 12)\,dx$

$= \left[\dfrac{x^3}{3} - 4x^2 + 12x \right]_0^2 = \dfrac{8}{3} - 16 + 24 = 10\tfrac{2}{3}.$

EXAMPLES XII. a.

1. Prove that the area bounded by the axis of x, the curve $y^2 = 4ax$, and the ordinate at the point (h, k) on the curve is $\dfrac{2hk}{3}$.

2. Find the area bounded by the axis of x, the curve $y^2 = 4a(x - 2a)$, and the ordinate at the point (h, k).

3. Find the area bounded by the axes of co-ordinates, the curve $x^2 = 4a(y - 2a)$, and the ordinate at the point (h, k).

4. Find the area of the ellipse $\dfrac{x^2}{a^2} + \dfrac{y^2}{b^2} = 1$.

5. Find the area between the curve $y = 2x - x^2$ and the axis of x.

6. Find the area between the curve $y = 5x - x^2 - 6$ and the axis of x.

7. Find the area between the curve $y^2 = 16x$ and the line $x = 8$.

8. Find the area between the curve $x^2 = 8y$, the axis of x, and the line $x = 6$.

9. Find the area enclosed in the 4th quadrant by the curve $y = x^3 - 5x^2 + 4x$ and the axis of x.

10. Find the area between the curve $x^2 = 4y$ and the line $y = 9$.

11. Find the area of a sector of a circle of radius a and angle θ.

12 On AA′, the major axis of an ellipse as diameter, describe a circle, and let PN, QM, ordinates of the circle, cut the ellipse at P′ and Q′ respectively, the points P and Q being very near to one another.

$$\frac{\text{The area of P′NMQ′}}{\text{the area of PNMQ}} = \frac{\text{P′N . MN}}{\text{PN . MN}} \text{ approx.}$$ This is true for all such elemental strips. From the area (πa^2) of the circle, deduce the area of the ellipse.

13. Find the area bounded by the curve $y = x^3 - 3x^2 + 5x$, the axis of x, and the line $x = 3$.

14. Find the area between the curve $y = \sin x$ and the axis of x, between the limits $x = 0$ and $x = \pi$, x being measured in radians.

15. Find the area between the curve $y = \sin x°$ and the axis of x, between the limits $x = 0°$, $x = 180°$.

16. Find the area between the curve $c^2 y = x^3$, the axis of x, and the ordinates corresponding to the abscissae a and b.

17. Find the area of the curve $y^2 = 8x$ cut off by the straight line $y = 2x$.

18. Draw the graph of $y = \dfrac{4}{x^2 + 1}$, and find the area between the curve and the axis of x.

19. Trace the curve $y = \dfrac{5}{x^2 + 4x + 5}$, and find the area between it and the axis of x. (For integration purposes let $x + 2 = u$.)

20. Find the area of the hyperbola $x^2 - y^2 = 9$ cut off by the straight line $x = 5$.

POLAR CO-ORDINATES.

Take two points P and Q on the curve and near to one another, and let their co-ordinates be (r, θ) and $(r + \Delta r, \theta + \Delta \theta)$ respectively. Also let ΔA denote the area OPQ.

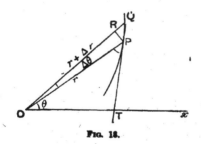

FIG. 18.

Then ΔA lies between $\dfrac{1}{2} r^2 \Delta \theta$ and $\dfrac{1}{2}(r + \Delta r)^2 \Delta \theta$.

$\therefore \dfrac{\Delta A}{\Delta \theta}$,, ,, $\dfrac{r^2}{2}$,, $\dfrac{1}{2}(r + \Delta r)^2$.

And in the limit $\dfrac{dA}{d\theta} = \dfrac{r^2}{2}$.

Hence $A = \displaystyle\int \dfrac{r^2}{2}\, d\theta$, between the proper limits.

EXAMPLE. *If* A *is the vertex of the parabola* $\dfrac{2a}{r} = 1 + \cos\theta$, *and* θ *is the angular co-ordinate of* P, *find the area of the sector* ASP, S *being the origin.*

A, the area required,

$$= \frac{1}{2}\int_0^\theta r^2 d\theta = 2a^2 \int_0^\theta \frac{d\theta}{(1 + \cos\theta)^2} = \frac{a^2}{2}\int_0^\theta \frac{d\theta}{\cos^4\dfrac{\theta}{2}}$$

$$= \frac{a^2}{2}\int_0^\theta \sec^4\frac{\theta}{2}\, d\theta = \frac{a^2}{2}\int_0^\theta \sec^2\frac{\theta}{2}\Big(1 + \tan^2\frac{\theta}{2}\Big)\, d\theta. \quad\text{........(1}$$

To integrate this expression, let $\tan\dfrac{\theta}{2} = u$, so that

$$\sec^2\frac{\theta}{2}\, d\theta = 2\,du.$$

Then $\dfrac{a^2}{2}\displaystyle\int \sec^2\frac{\theta}{2}\Big(1 + \tan^2\frac{\theta}{2}\Big)\, d\theta = a^2 \int (1 + u^2)\, du = a^2\Big(u + \frac{u^3}{3}\Big)$

$$= \frac{a^2}{3} u(3 + u^2) = \frac{a^2}{3}\tan\frac{\theta}{2}\Big(3 + \tan^2\frac{\theta}{2}\Big).$$

$\therefore$ from (1), $A = \dfrac{a^2}{3}\tan\dfrac{\theta}{2}\Big(3 + \tan^2\dfrac{\theta}{2}\Big).$

EXAMPLES XII. b.

1. Draw the curve $r^2 = a^2\cos 2\theta$, and find its area.
2. Draw the curve $r = a(1 + \cos\theta)$, paying particular attention to its shape near the origin, and find its area.
3. In the circle $r = 2a\cos\theta$, find by calculus methods the area enclosed by the curve and the lines $\theta = 0$, $\theta = \dfrac{\pi}{4}$. Check your result by mensuration.
4. Draw the curve $r = 4(1 - \cos\theta)$, and find its area.
5. If C is the centre of an ellipse, A a vertex, and P a point on it such that the angle ACP $= \theta$, find the area of the sector ACP.

$\Big[$Use C as origin, CA as initial line, and for integration purposes let $\dfrac{a\tan\theta}{b} = u.\Big]$

To find the curved surface of a right circular cone.

By placing the cone with a slant side on a plane, and rolling the cone on the plane, keeping the vertex fixed, we see that the surface of the cone can be unwrapped into a *plane* surface.

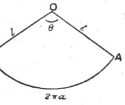

It becomes a sector of a circle whose arc is the circumference of the base of the cone, and whose radius is the slant side of the cone.

Fig. 19.

∴ if $a=$ the radius of the base,

$l=$ the length of the slant side, and $\theta=$ the angle of the sector,

the curved surface of the cone $=\frac{1}{2}l^2\theta \qquad \left(\dfrac{r^2\theta}{2}\right)$

$$=\tfrac{1}{2}l^2 \cdot \frac{2\pi a}{l} \qquad (\text{arc}=a\theta)$$

$$=\pi a l$$

$$=\frac{\pi a^2}{\sin \alpha}, \text{ if } \alpha \text{ is the semi-vertical} \atop \text{angle of the cone,}$$

$$=\pi l^2\sin \alpha.$$

Or the following method might be used:

Let PQ be any small portion of the circumference of the base, and O the vertex.

Draw ON perpendicular to PQ.

The small portion OPQ of the surface is approximately a triangle, and hence its area may be taken to be $\frac{1}{2}$PQ.ON, *i.e.* $\frac{1}{2}$PQ.l.

∴ the total surface $=$ the sum of all the small areas like $\frac{1}{2}$PQ.l

Fig. 20.

$$=\frac{l}{2}\times \qquad \text{,,} \qquad \text{,,} \qquad \text{arcs like PQ}$$

$$=\frac{l}{2}\times \text{ the circumference of the base}$$

$$=\pi a l.$$

To find the area of the curved surface of the belt of a cone.

If PRSQ is the belt, and NP, MQ the radii of its plane ends,

$$\text{its area} = \frac{\pi QM^2}{\sin a} - \frac{\pi PN^2}{\sin a} = \frac{\pi(QM - PN)(QM + PN)}{\sin a}$$

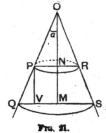

(a is the semi-vertical angle of the cone)

$$= \frac{\pi QV \cdot (QM + PN)}{\sin a}, \text{ if PV is perp}^r \text{ to QM}$$

$$= \pi PQ(QM + PN).$$

Hence if a' is the mean of the radii of the two ends,

Fig. 21.

the surface of the belt $= 2\pi a' \times PQ$

To find the curved surface of (1) *a sphere of radius a,* (2) *a belt of the sphere,* (3) *a cap of the sphere.*

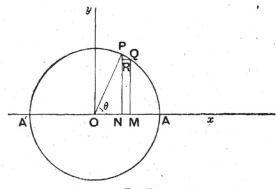

Fig. 22.

Suppose the sphere to be generated by the revolution of the semicircle APA′ about Ox, and let ΔS be the element of surface generated by the small arc PQ.

[The construction can be gathered from the figure.]

Then $\quad \Delta S = 2\pi PN \cdot PQ$

$$= 2\pi \cdot PN \cdot \frac{QR}{\sin \theta}, \text{ for } \angle RPQ = 90° - \angle OPN = \theta,$$

$$= 2\pi a \cdot \Delta x.$$

$$\therefore \frac{dS}{dx} = 2\pi a.$$

∴ (1) S the area of the sphere $= 2\int_0^a 2\pi a\, dx = 4\pi a^2$.

(2) If h is the perpendicular distance between the ends of the belt,

the surface of the belt $= \int_x^{x+h} 2\pi a\, dx = 2\pi a h$.

(3) If h is the height of the cap,

the surface of the cap $= \int_{a-h}^{a} 2\pi a\, dx = 2\pi a h$.

LENGTHS OF ARCS.

If $P(x, y)$ and $Q(x + \Delta x, y + \Delta y)$ are two points on a curve, near to one another, and Δs is the length of the small arc PQ, $(\Delta x)^2 + (\Delta y)^2 = PQ^2 = (\Delta s)^2$ in the limit.

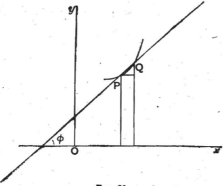

Fig. 23.

Or, expressing this in differentials, $ds^2 = dx^2 + dy^2$.

$$\therefore \frac{ds}{dx} = \sqrt{1 + \left(\frac{dy}{dx}\right)^2} \quad \text{and} \quad s = \int \sqrt{1 + \left(\frac{dy}{dx}\right)^2} \cdot dx.$$

We also see that if the tangent makes an angle ϕ with the axis of x,

$$\frac{dx}{ds} = \cos\phi, \quad \text{and} \quad \frac{dy}{ds} = \sin\phi.$$

If polar co-ordinates are used, from the accompanying diagram,

$$PR^2 + QR^2 = PQ^2.$$

∴ in the limit, $\quad \overline{rd\theta}|^2 + dr^2 = ds^2.$

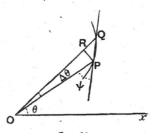

Fig. 24.

$$\therefore \frac{ds}{dr} = \sqrt{1 + \left(\frac{rd\theta}{dr}\right)^2} \quad \text{and} \quad s = \int \sqrt{1 + \left(\frac{rd\theta}{dr}\right)^2} \cdot dr.$$

Or, $\quad \dfrac{ds}{d\theta} = \sqrt{r^2 + \left(\dfrac{dr}{d\theta}\right)^2} \quad \text{and} \quad s = \int \sqrt{r^2 + \left(\dfrac{dr}{d\theta}\right)^2} \cdot d\theta.$

N.B.—If ψ is the angle between the tangent and radius vector, $PR = r\sin\Delta\theta = rd\theta$ approx.

$$\sin\psi = \text{the limiting value of } \frac{PR}{PQ}$$

$$= \quad \text{,,} \quad \text{,,} \quad \frac{r\Delta\theta}{\Delta s} = \frac{rd\theta}{ds}$$

$$\cos\psi = \quad \text{,,} \quad \text{,,} \quad \frac{QR}{QP}$$

$$= \quad \text{,,} \quad \text{,,} \quad \frac{\Delta r}{\Delta s} = \frac{dr}{ds},$$

and $\quad \tan\psi = \quad \text{,,} \quad \text{,,} \quad \dfrac{PR}{QR} = \dfrac{rd\theta}{dr}.$

EXAMPLES XII. c.

1. A quadrant ABCD of a circle, of radius 1 ft., is trisected at B and C. Find the area of the belt generated by the revolution of the arc BC about the radius OD.

2. An arc AB of a circle, radius 4 inches, subtends an angle of 60° at the centre O. Find the area (to the nearest tenth of a sq. in.) of the total surface of the solid generated by the revolution of the sector OAB about the radius OA.

3. $x = a(\theta + \sin \theta)$, $y = a(1 - \cos \theta)$ are the equations of a cycloid. Find the length of the arc between the points given by $\theta = 0$, $\theta = \pi$.

$$\left[\text{Here } ds^2 = dx^2 + dy^2 = a^2[(1 + \cos \theta)^2 + \sin^2\theta]d\theta^2, \text{ whence } ds = 2a \cos \frac{\theta}{2}d\theta. \right]$$

4. Draw the curve $r = a(1 - \cos \theta)$, and find its length.

VOLUMES.

If AB *is a known curve,* BD *the ordinate at* B, *and the area* AODB *revolves about the axis of* x, *it is required to find the volume generated.*

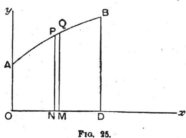

FIG. 25.

Draw ordinates PN, QM at two points P and Q very near to one another. Let (x, y) be the co-ors. of P and $(x + \Delta x, y + \Delta y)$ those of Q.

The area of the circle generated by the revolution of PN
$= \pi . PN^2$.

$\therefore$ the volume of the lamina generated by the revolution of
the strip PNMQ $= \pi PN^2 . MN$, approx.,

$$= \pi y^2 . \Delta x.$$

Hence, if V is the volume required,

$$\Delta V = \pi y^2 \Delta x,$$

and proceeding to the limit,

$$\frac{dV}{dx} = \pi y^2.$$

$\therefore$ if h is the abscissa of the point B,

V, the volume required $= \int_0^h \pi y^2 dx$.

NOTE. The volume of the lamina generated by the revolution of
PNQM lies *between* $\pi y^2 \Delta x$ and $\pi (y + \Delta y)^2 \Delta x$.
Hence, in taking $\Delta V = \pi y^2 \Delta x$, we only neglect products of Δx, Δy, and
their powers.

EXAMPLES XII. d.

1. Find the volume of a right circular cone of height h,
   a being the radius of its base.
   [Take the vertex at the origin, and the axis of the cone
   as axis of x.]

2. Find the volume of a sphere of radius a.
   [Consider the sphere as generated by the revolution of a
   semi-circle about its diameter.]

3. Find the volume of a segment of height h of a sphere
   whose radius is a.

4. The arc of the parabola $y^2 = 4ax$ between the origin and
   the point (h, k) revolves about the axis of x: find the
   volume generated.

5. The straight line $2y = 3x + 1$ revolves about the axis of x.
   Find the volume generated between the limits $x = 3$,
   $x = 1$.

6. A circle, centre $(4, 0)$, radius 4, revolves about the axis of x. Find the volume generated between the limits $x = 2$, $x = 0$.

7. The curve $y = e^{2x}$ revolves about the axis of x. Find the volume generated between the values $x = 4$, $x = 0$.

8. In the parabola $y^2 = -4a(x - h)$, the part in the first quadrant revolves about Ox. Find the volume generated.

CHAPTER XIII.

CENTRE OF GRAVITY. CENTRE OF PRESSURE. WORK.

In finding the C.G. of a given area or volume, we use the Principle of Moments.

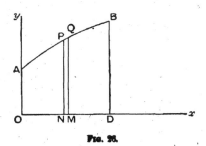

Fig. 26.

If A is the whole area, ρ its density, $(\bar{x}, \bar{y})$ the co-ors. of its C.G., and PNMQ an elemental strip, then $\rho A\bar{x}$ is the moment of the whole area about Oy as axis.

The mass of the area PNMQ $= \rho y \Delta x$,

and its moment about Oy as axis $= \rho xy \Delta x$.

$\therefore \rho A\bar{x} = $ the sum of all such elements of moment as $\rho xy \Delta x$.

i.e. $\rho A\bar{x} = \rho \int xy\, dx$ between the proper limits.

In the same way, $\rho A\bar{y} = \dfrac{\rho}{2} \int y^2 dx$ between the proper limits, for the distance of the C.G. of PNMQ from Ox $= \dfrac{y}{2}$.

When possible we take the axis of x along a line of symmetry.

To find the C.G. of a sector of a circle of radius a, 2a being the angle at the centre.

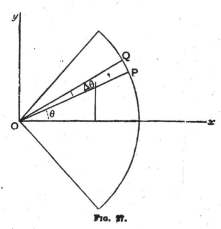

FIG. 27.

Take the axis of x along the line bisecting the sector.

Let OPQ be an element of area, $\angle POx = \theta$, $\angle POQ = \Delta\theta$.

By symmetry, the C.G. lies in Ox; let $\bar{x}$ be its distance from O, and let (x, y) be the co-ors. of P.

The area of $OPQ = \frac{1}{2}a^2\Delta\theta$.

The distance of its C.G. from $O = \frac{2}{3}a$, for OPQ is approximately a $\triangle$.

$\therefore$ the distance of its C.G. from $Oy = \frac{2}{3}x$.

$\therefore$ its moment about Oy as axis $= \rho\dfrac{a^2\Delta\theta}{2} \times \dfrac{2x}{3} = \rho\dfrac{a^3}{3}\cos\theta\,\Delta\theta$.

The area of the whole sector $= \rho\dfrac{a^2}{2} \cdot 2a$.

$\therefore$ by the Principle of Moments,

$$\rho\frac{a^2}{2} \cdot 2a\bar{x} = \rho\frac{a^3}{3}\int_{-a}^{a}\cos\theta\,d\theta.$$

$$\therefore a\bar{x} = \frac{a}{3}\left[\sin\theta\right]_{-a}^{a} = \frac{2a}{3}\sin a.$$

$$\therefore \bar{x} = \frac{2a}{3}\frac{\sin a}{a}.$$

GULDIN'S OR PAPPUS'S THEOREMS.

I. *If a plane curve rotates about any external axis, situated in its plane, the area of the surface generated is equal to the product of the length of the rotating curve and the length of the path described by its centre of gravity.*

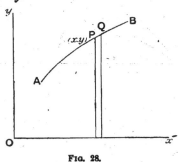

FIG. 28.

If the curve AB, of length s, rotates about Ox, the surface generated by the element $PQ(\Delta s) = 2\pi y \Delta s$.

∴ the total surface generated by $AB = 2\pi \int y \, ds$ between the proper limits.

But if $\bar{y}$ is the ordinate of the centre of gravity of the curve by the principle of moments, $\bar{y} \cdot s = \int y \, ds$.

∴ the surface generated $= 2\pi \bar{y} \cdot s$.   Q.E.D.

N.B.—The centre of gravity of the arc AB is used here, not the centre of gravity of an area.

II. *If a plane curve rotates about any external axis, situated in its plane, the volume of the solid generated is equal to the product of the area of the rotating curve and the length of the path described by its centre of gravity.*

If the curve, area A, rotates about Ox, the volume generated by an element of area (ΔA) corresponding to the point (x, y)

$$= 2\pi y \Delta A.$$

∴ the total volume generated $= 2\pi \int \bar{y} \, dA$ ....(1)

But, by the principle of moments, if $\bar{y}$ is the ordinate of the centre of gravity of the rotating curve,

$$\bar{y}A = \int y \, dA.$$

∴ from (1), the volume generated $= 2\pi\bar{y} \cdot A.$ Q.E.D.

EXAMPLES XIII. a.

Find the C.G. of :

1. A semi-circle.

2. A quadrant of a circle.

3. A right circular cone.

4. An arc of a circle, $2a$ being the angle at the centre.

5. The portion of the parabola $y^2 = 4ax$ bounded by the curve and the double ordinate at the point (h, k).

6. A quadrant of an ellipse.

7. A semi-ellipse.

8. The portion of the parabola $y^2 = 4ax$ bounded by the curve, the ordinate at the point (h, k), and the axis of x.

9. A curve, whose equation is $xy = a$ constant, passes through the point $(4, 5)$. Find the area between the curve, the axis of x, and the two ordinates $x = 4$, $x = 12$. Also find the centre of gravity of this area.

10. A curtain ring has an internal diameter of $3\frac{1}{2}$ inches, and its cross-section is a circle of diameter $\frac{3}{4}$ of an inch.

Find   (i) its surface ;
       (ii) its volume.

To find the total pressure on an isosceles triangle of height k immersed in a fluid of density ρ, so that its vertex A is at a depth h, and its base BC(a) is horizontal.

Let AD be perp^r to the base, and take a very narrow strip EF of the △ parallel to the base and at a distance x from A.

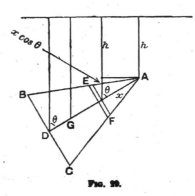

Fic. 29.

Let AD make an ∠θ with the vertical.

$$\frac{EF}{BC} = \frac{x}{k} \qquad \therefore \ EF = \frac{ax}{k}.$$

The pressure on $EF = \rho EF . \Delta x(h + x \cos \theta)$ with the usual notation ;

i.e. if P is the total pressure, $\Delta P = \dfrac{\rho ax}{k}(h + x \cos \theta)\Delta x.$

$\therefore$ in the limit, $\quad \dfrac{dP}{dx} = \dfrac{\rho ax}{k}(h + x \cos \theta).$

$$\therefore \ P = \frac{\rho a}{k}\int_0^k (hx + x^2 \cos \theta)\, dx$$

$$= \frac{\rho a}{k}\left[\frac{hx^2}{2} + \frac{x^3 \cos \theta}{3}\right]_0^k$$

$$= \rho ak\left[\frac{h}{2} + \frac{k \cos \theta}{3}\right]$$

$$= \frac{1}{2}\rho ak\left(h + \frac{2k}{3}\cos \theta\right).$$

N.B.—The area of the $\triangle = \frac{1}{2} ak$. The depth of its C.G. is $h + \frac{2}{3}$ AD cos θ, *i.e.* $h + \frac{2k}{3}$ cos θ.

$\therefore$ P $= \rho \times$ the area $\times$ the depth of its C.G.

$=$ the wt. of a column of fluid whose base is horizontal and equal to the area of the $\triangle$, and whose height is equal to the depth of the C.G.

Definition. The point in a surface at which the resultant fluid pressure acts is called the **Centre of Pressure.**

To find the centre of pressure of the triangle in the preceding example.

The moment about A of the pressure on the strip EF

$$= \frac{\rho a x^2}{k}(h + x \cos \theta) \Delta x.$$

$\therefore$ the sum of all such moments $= \frac{\rho a}{k} \int_0^k (x^2 h + x^3 \cos \theta) dx.$

$\therefore$ if $\bar{x}$ is the distance of the centre of pressure from A,

$$P\bar{x} = \frac{\rho a}{k} \int_0^k (x^2 h + x^3 \cos \theta) dx = \frac{\rho a}{k} \left[\frac{x^3 h}{3} + \frac{x^4 \cos \theta}{4} \right]_0^k$$

$$= \frac{\rho a}{k} \left[\frac{k^3 h}{3} + \frac{k^4 \cos \theta}{4} \right].$$

$$\therefore \bar{x} = \frac{\frac{\rho a k^2}{12}(4h + 3k \cos \theta)}{\frac{\rho a k}{6}(3h + 2k \cos \theta)} = \frac{k(4h + 3k \cos \theta)}{2(3h + 2k \cos \theta)}.$$

EXAMPLES XIII. b.

1. Find (1) the total pressure, (2) the centre of pressure in the case of an isosceles $\triangle ABC$, immersed in a fluid, the vertex A being in the surface, its base horizontal, and the plane of the $\triangle$ vertical.

2. Find (1) the total pressure, (2) the centre of pressure of the $\triangle$ in Example 1 when the vertex A is at a depth h.

3. Find (1) the total pressure, (2) the centre of pressure in the case of a rectangle immersed in a fluid with one side a in the surface and the other side b vertical.

4. Find (1) the total pressure, (2) the centre of pressure in the rectangle in Example 3 when the upper horizontal side a is at a depth h.

5. An isosceles $\triangle ABC$ is immersed in a fluid with its base BC in the surface and its plane vertical. Find the total pressure on the $\triangle$ and its centre of pressure.

6. Find the total pressure and the centre of pressure in the $\triangle$ of Example 5 when the base is at a depth h.

7. A rectangle is immersed in a fluid with one side a in the surface, and the other side b inclined at an angle a. to the horizontal. Find the total pressure on the rectangle and the position of the centre of pressure.

8. In the rectangle of the above question, find the total pressure and the centre of pressure when the side a is immersed at a depth k.

9. A cubical box of edge 1 ft., filled with water, has a close fitting iron lid, half an inch thick, fixed by smooth hinges to one edge; and the box is tilted through an angle θ about the opposite edge of the base. Find the total pressure and the centre of pressure of the water on the lid, and determine the angle at which the water will begin to escape. (Take the density of iron as 7·6 times that of water and neglect the atmospheric pressure on the lid.)

10. A rectangular plate used in connection with a weir is immersed vertically; its upper and lower edges are horizontal, and are immersed at x cm. and y cm. respectively from the surface of the water. Determine an expression in terms of x and y only for the depth of the centre of pressure, and given that the length of the horizontal sides is a cm., and the length of the other two sides is b cm., find an expression for the total thrust upon this plate.

Work done by a variable force.

If ΔW is the element of work done over a small space Δs, measured in the direction of P the moving force,

$$\Delta W = P\Delta s; \quad \therefore \text{ in the limit, } \frac{dW}{ds} = P.$$

$$\therefore W = \int P\,ds \text{ between the proper limits.}$$

Energy. If E is the energy of a body under the action of a variable force P,

$$\Delta E = \Delta W = P\Delta s; \quad \therefore \text{ in the limit, } \frac{dE}{ds} = P.$$

$$\therefore E = \int P\,ds.$$

Hooke's Law. *The tension of an elastic string varies as its extension beyond its unstretched length.*

This is usually expressed thus :

If l is the unstretched length of a string, and P the force required to stretch it to length l',

$$P = \lambda \frac{l' - l}{l}, \text{ where } \lambda \text{ is constant.}$$

Spiral Springs. *In compressing or extending spiral springs, the force required in any position may be taken as proportional to the amount of compression or extension.*

EXAMPLES XIII. c.

1. If a force of 10 lb. is required to stretch an elastic string through one inch, find the work done in stretching it through 4 inches.

2. If a force of 48 lb. will compress a spiral spring through one inch, find the work done in compressing it through 5 inches.

3. In an air-compressing plant, 12 c. ft. of air, at a pressure
of 15 lb. per sq. in. absolute, are drawn into the compression cylinder and compressed by the piston into a
volume of 3 c. ft.   During the compression the air is
kept at a constant temperature.

Find (a) the absolute pressure of the air at the end of
the compression in lb. per sq. in.; (b) the number of
ft.-lb. of work (to the nearest 100) which have been
done in compressing the air.

CHAPTER XIV.

MOMENTS OF INERTIA.

IF all the particles of a body of mass m are moving in a straight line with velocity v, the kinetic energy of the body is $\frac{1}{2}mv^2$. If, however, the body is rotating in any way, we cannot write down its kinetic energy, for different particles of the body are moving with different velocities.

Moments of Inertia enable us to determine the kinetic energy of a body revolving about an axis.

If m is the mass of a particle of the body and r is the distance of this particle from the axis of rotation, the sum of mr^2 (taken for particles throughout the body) is called the Moment of Inertia of the body about this axis of rotation.

This may be written $\Sigma(mr^2)$, or with the notation of the calculus, $\int r^2 dm$, where r is the distance of the element of mass, dm, from the axis of rotation.

If $\int r^2 dm = Mk^2$, where M is the mass of the body, k is called the *Radius of Gyration*.

The moment of inertia of a body is often denoted by the letter I. If ω is the angular velocity of the body about the axis of rotation, the kinetic energy of any element dm is $\frac{1}{2}dm(r\omega)^2$, for $r\omega$ is its linear velocity.

$\therefore$ the K.E. of the whole body $= \int \frac{1}{2}r^2\omega^2 dm = \frac{\omega^2}{2}\int r^2 dm = \frac{\omega^2 I}{2}.$

EXAMPLE 1.   *A thin uniform rod of length a and density ρ rotates with angular velocity ω in a plane about an axis at one end, the axis being at right angles to the rod. Find the moment of inertia of the rod and the work it is capable of doing in being brought to rest.*

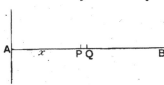

FIG. 30.

If AB is the rod, and PQ(Δx) any element of length, AP being equal to x, the moment of inertia of PQ about the axis through A $= \rho\Delta x \cdot x^2$, for $\rho\Delta x$ is the mass of PQ.

∴ the moment of inertia of the whole rod

$$= \int_0^a \rho x^2 dx = \rho \left[\frac{x^3}{3}\right]_0^a = \frac{\rho a^3}{3}.$$

If m is the mass of the rod, $m = \rho a$,

and the moment of inertia $= \dfrac{ma^2}{3}$.

The work required = its kinetic energy

$$= \frac{\omega^2 I}{2} = \frac{ma^2\omega^2}{6}.$$

EXAMPLE 2.   *To find the moment of inertia of a triangle about one side.*

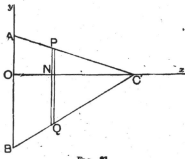

FIG. 31.

Let ABC be the △, and draw CO perpendicular to the base AB.   Take OC and OA as axes of x and y respectively.

Let the altitude of the △, OC, be h.

Take PQ an elemental strip of the $\triangle$ parallel to Oy.

If Δx is its width, x being the abscissa, ON, of the point P,

$$\frac{PQ}{AB} = \frac{CN}{OC}, \quad i.e. \quad PQ = \frac{c(h-x)}{h}.$$

$\therefore$ the area of the strip $= \dfrac{c(h-x)}{h} \Delta x$,

and since all parts of the strip are equidistant from Oy, its moment of inertia about $Oy = m\dfrac{c(h-x)x^2\Delta x}{h}$, where m is the mass of unit area.

$\therefore$ the moment of inertia of the whole $\triangle = m\dfrac{c}{h}\displaystyle\int_0^h (h-x)x^2 dx$

$$= m\frac{c}{h}\left[\frac{x^3h}{3} - \frac{x^4}{4}\right]_0^h = m\frac{ch^4}{h}\left(\frac{1}{3} - \frac{1}{4}\right) = m\frac{ch^3}{12}$$

$$= m\frac{A \cdot h^2}{6}, \text{ if A is the area of the } \triangle, \text{ for } A = \frac{ch}{2}.$$

The following theorems are useful in determining moments of inertia:

If $\bar{x}, \bar{y}, \bar{z}$ are the co-ordinates of the centre of gravity of a body, and m is any particle of the body whose co-ordinates are x, y, z, then

$$\bar{x} = \frac{\Sigma(mx)}{\Sigma(m)}, \quad \bar{y} = \frac{\Sigma(my)}{\Sigma(m)}, \quad \bar{z} = \frac{\Sigma(mz)}{\Sigma(m)}.$$

We have used this property in the chapter on Centre of Gravity.

If Ox, Oy, Oz are axes mutually at right angles to one another, and I_x, I_y, I_z are the moments of inertia about Ox, Oy, Oz respectively of any plane lamina in the plane xOy,

$$I_x + I_y = I_z.$$

If (x, y) are the co-ordinates of any point in the plane xOy, and r its distance from the origin O, $x^2 + y^2 = r^2$.

$$\therefore \Sigma(mx^2) + \Sigma(my^2) = \Sigma(mr^2) \dots\dots\dots\dots(1)$$

But $I_x = \Sigma(my^2)$, $I_y = \Sigma(mx^2)$; and r is the perpendicular distance of the point (x, y) from Oz; $\therefore \Sigma(mr^2) = I_z$.

$\therefore$ from (1), $\qquad I_x + I_y = I_z$.

If I *is the moment of inertia of any body of mass* M *about any axis* O_z, I_g *its moment of inertia about a parallel axis through* G *its centre of gravity, and d the distance between the two axes,*

$$I = I_g + Md^2.$$

Let m be the mass of any particle of the body. Take Oz as axis of z, Ox, Oy being the axes of x and y.

Let $(\bar{x}, \bar{y}, \bar{z})$ be the centre of gravity of the body, (x, y, z) the co-ordinates of m, and (x', y', z') the co-ordinates of m referred to parallel axes through G, so that

$$x = \bar{x} + x', \quad y = \bar{y} + y', \quad z = \bar{z} + z'.$$

Then I, the moment of inertia about $Oz = \Sigma[m(x^2 + y^2)]$

$$= \Sigma[m\{(\bar{x} + x')^2 + (\bar{y} + y')^2\}]$$

$$= \Sigma[m(\bar{x}^2 + \bar{y}^2)] + 2\bar{x}\Sigma(mx') + 2\bar{y}\Sigma(my') + \Sigma[m(x'^2 + y'^2)]....(1)$$

Now $\bar{x}^2 + \bar{y}^2 = d^2$, so that $\Sigma[m(\bar{x}^2 + \bar{y}^2)] = Md^2$.

Also, $\Sigma(mx') = 0$ and $\Sigma(my') = 0$, for $\dfrac{\Sigma(mx')}{\Sigma(m)}$, $\dfrac{\Sigma(my')}{\Sigma(m)}$ are the xy co-ordinates of the centre of gravity of the body referred to axes through G.

Also $\Sigma m(x'^2 + y'^2) = I_g$, as defined in the enunciation.

$\therefore$ from (1), $\qquad I = Md^2 + I_g.$ $\qquad\qquad$ Q.E.D.

EXAMPLE. *The moment of inertia of an ellipse of mass* M *about its minor axis is* $\dfrac{Ma^2}{4}$.

Take $\dfrac{x^2}{a^2} + \dfrac{y^2}{b^2} = 1$ for the equation of the ellipse, and let PQ be an elemental strip parallel to Oy.

The moment of inertia is

$$2m\int_{-a}^{a} x^2 y \, dx = 2m\frac{b}{a}\int_{-a}^{a} x^2\sqrt{a^2 - x^2}\,dx,$$

where m is the mass of unit area.

Taking $x = a \sin \theta$, this becomes

$$2ma^3b \int_{-\frac{\pi}{2}}^{\frac{\pi}{2}} \sin^2\theta \cos^2\theta \, d\theta = \frac{ma^3b}{4} \int_{-\frac{\pi}{2}}^{\frac{\pi}{2}} (1 - \cos 4\theta) \, d\theta$$

$$= \frac{m\pi a^3 b}{4} = \frac{Ma^2}{4}.$$

Routh gives the following rule, which may serve as a check. The moment of inertia about an axis of symmetry

$$= \text{mass} \frac{\text{(the sum of the squares of perpendicular semi-axes)}}{3, \, 4, \, \text{or} \, 5}.$$

The denominator is 3, 4, or 5 according as the body is rectangular, elliptical, or ellipsoidal.

E.g., In taking the moment of inertia of a circle (which is elliptical), of radius a, about a diameter, the perpendicular semi-axis in its plane is the radius a;

the semi-axis perpendicular to its plane is zero.

∴ the moment of inertia required is $\frac{Ma^2}{4}$.

The denominator 5 will be taken in the case of a sphere, which is ellipsoidal.

The time of a complete oscillation (swing-swang) of a simple pendulum is $2\pi \sqrt{\frac{l}{g}}$, where l is the length of the pendulum.

In this case we neglect the mass of the supporting string or wire, and we treat the bob as a particle. But with an ordinary pendulum such as that used in a clock, the wire is of appreciable mass, and the bob of appreciable size.

If t is the time of a beat of a pendulum of any shape, then the simple pendulum which oscillates in the same time, t, is called the **simple equivalent pendulum**.

The simple equivalent pendulum of a mass M oscillating about a horizontal axis.

Let h be the distance of the centre of gravity of the body from Ox, the axis of rotation, and ω the angular velocity of the body at any instant.

Considering a particle of mass m at a distance r from Ox, its velocity perpendicular to the direction of r is $r\omega$.

$\therefore$ its acceleration in this direction is $\dfrac{d}{dt}(r\omega) = r\dfrac{d\omega}{dt}$, for r is constant.

The acceleration of the particle in the direction of r is $r\omega^2$.

Hence, the forces acting on the particle are :

$$mr\dfrac{d\omega}{dt} \text{ perp}^{r} \text{ to the direction of } r,$$

and $mr\omega^2$ in the direction of r.

The moment of the first force about O$x = mr^2\dfrac{d\omega}{dt}$.

    ,,   ,,   ,,   second   ,,   ,, $= 0$.

$\therefore$ the sum of the moments of the particles comprising M about Ox

$$= \Sigma(mr^2)\dfrac{d\omega}{dt}$$

$$= Mk^2\dfrac{d\omega}{dt}, \text{ where } k \text{ is the radius of gyration. } ...(1)$$

If θ is the inclination of h to the vertical,

the moment of M about O $= Mgh\sin\theta$.

$\therefore$ from (1), $Mk^2\dfrac{d\omega}{dt} = -Mgh\sin\theta,$

or, $k^2\dfrac{d\omega}{dt} = -gh\sin\theta.$

The similar equation of motion for a particle of any mass suspended by a string of length l is

$$l^2\dfrac{d\omega}{dt} = -gl\sin\theta,$$

and hence we see that the angular motions of the body and the string will be the same if $\dfrac{k^2}{l^2} = \dfrac{h}{l}$, i.e. if $l = \dfrac{k^2}{h}$.

$\therefore \dfrac{k^2}{h}$ is the length of the simple equivalent pendulum.

EXAMPLES XIV.

[The student should check his results by Routh's rule.]

1. Prove that $\dfrac{Ma^2}{2}$ is the moment of inertia of a circular plate, of radius a and mass M, about an axis perpendicular to the plate and passing through its centre. If the plate revolves about this axis with uniform angular velocity ω, deduce its kinetic energy.

2. A thin uniform rectangle of mass m, sides a and b, rotates with uniform angular velocity ω about one side a: find its moment of inertia about this side and its kinetic energy.

3. In the plate of Example 1, use the fact that $I_x + I_y = I_z$ to find the moment of inertia of the plate about a diameter.

4. Find the moment of inertia of the rectangle in Example 2 about an axis through its centre of inertia parallel to the sides a, (1) by deduction from the result of Example 2, using $I = I_g + Md^2$; (2) independently.

5. A uniform metal rod AB, of length a and mass m, has a piece CBD of like material soldered to it at right angles at B. Find the moment of inertia of the two rods about an axis through A parallel to CBD.

6. Find the moment of inertia of a triangle, of mass m and height h, about an axis through its vertex and parallel to its base.

7. Find the moment of inertia of a sphere about a diameter. [Take a thin lamina at right angles to Ox, and use the result of Example 1 to write down its moment of inertia about Ox. Then integrate.]

8. A rod 100 cm. long is suspended by one end and oscillates freely through a small angle under the influence of gravity. Find the length of the simple equivalent pendulum and the period of vibration.

9. Find the moment of inertia of a uniform circular disc, of radius r and mass M, about a tangent.

    Also find the time of complete oscillation of the disc when it swings about this line through a small angle.

10. Find the moment of inertia of a uniform circular disc, of radius r and mass M, about a line at right angles to its plane and through a point in its circumference. Also find the period of vibration when the disc oscillates about this line through a small angle.

CHAPTER XV.

CATENARY. CYCLOID. LOGARITHMIC OR EQUI-ANGULAR SPIRAL.

Catenary. If a uniform heavy inelastic string is suspended from two fixed points, it hangs in a curve (catenary) whose equation, referred to horizontal and vertical axes, is

$$y = \frac{a}{2}\left(e^{\frac{x}{a}} + e^{-\frac{x}{a}}\right).$$

[The proof of this equation will be found on p. 122.]

EXAMPLES XV.

1. Find the co-ordinates of the point A where this curve cuts the axis of y, and prove that this point is the lowest point on the curve.

2. If the tangent at the point (x, y) makes an angle ϕ with Ox, find the value of $\tan \phi$, and prove that $y = a \sec \phi$.

3. If PN is the ordinate at P, and NK is drawn perpendicular to the tangent at P, prove that NK is constant for all positions of P.

4. If the length of the arc AP $= s$, A being the lowest point, use the value of $\sec \phi$ found in Example 2 to prove that $s = a \tan \phi$. $\left(\text{Use the equation } \dfrac{dx}{ds} = \cos \phi.\right)$

5. If the normal at P meets the axis of x at G, prove that $a \cdot PG = y^2$.

6. Write down the value of $\tan \phi$ from the equation of the curve, and prove that $\frac{dx}{d\phi} = a \sec \phi$.

7. Prove that the area between the curve, the axes of co-ordinates, and the ordinate $PN = a^2 \tan \phi = 2 \triangle PKN$. (See Example 3 above.)

8. Prove that $PK = s$. (See Examples 3 and 4 above.)

9. Prove that the tension at any point varies as the ordinate at that point.

To find the equation of a catenary.

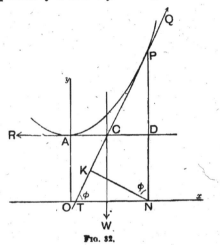

Fig. 32.

Consider the equilibrium of the portion AP, A being the lowest point.

The forces acting on it are
        (1) its own weight W,
        (2) R the tension at A, horizontally,
        (3) Q   ,,   P along the tangent.

Let the arc $AP = s$, and let a be the length of string whose weight is R, so that we may take $W = ks$ and $R = ka$.

Measure AO vertically downwards and equal to a.

Take OA as axis of y, and Ox, perpendicular to OA in the plane of the string, as axis of x.

Let the tangent PT make an angle ϕ with Ox, and let the ordinate PN meet the tangent at A in D.

Then, by the Triangle of Forces, $\dfrac{W}{R} = \dfrac{PD}{CD}$, ($\triangle$ PCD),

$$i.e. \quad \frac{s}{a} = \tan \phi = \frac{dy}{dx} \dots\dots\dots\dots\dots\dots(1)$$

$$ds^2 = dx^2 + dy^2 ; \quad \therefore \left(\frac{ds}{dy}\right)^2 = 1 + \left(\frac{dx}{dy}\right)^2 = 1 + \frac{a^2}{s^2}$$

and
$$\frac{dy}{ds} = \frac{s}{\sqrt{a^2 + s^2}}.$$

$\therefore$ by integration, $y = \sqrt{a^2 + s^2} + c$, where c is constant.
But when $y = a$, $s = 0$; $\therefore$ $c = 0$.
$$\therefore y = \sqrt{a^2 + s^2} \quad \text{and} \quad s^2 = y^2 - a^2.$$

Again, from (1), $\quad \dfrac{dx}{dy} = \dfrac{a}{s} = \dfrac{a}{\sqrt{y^2 - a^2}}.$

$$\therefore x = \int \frac{a}{\sqrt{y^2 - a^2}} dy = a \log (y + \sqrt{y^2 - a^2}) + c'.$$

But when $x = 0$, $y = a$; $\therefore$ $c' = -a \log a$.
$$\therefore x = a \log \left(\frac{y + \sqrt{y^2 - a^2}}{a}\right).$$

This is the equation of the curve, but it can be arranged in a more convenient form.

$$e^{\frac{x}{a}} = \frac{y + \sqrt{y^2 - a^2}}{a} \quad\dots\dots\dots\dots\dots\dots\dots(2)$$

$$= \frac{y^2 - (y^2 - a^2)}{a(y - \sqrt{y^2 - a^2})} = \frac{a}{y - \sqrt{y^2 - a^2}}.$$

$$\therefore e^{-\frac{x}{a}} = \frac{y - \sqrt{y^2 - a^2}}{a}.$$

$\therefore$ from (2), by addition,

$$e^{\frac{x}{a}} + e^{-\frac{x}{a}} = \frac{2y}{a} \quad \text{or} \quad y = \frac{a}{2}\left(e^{\frac{x}{a}} + e^{-\frac{x}{a}}\right).$$

Cycloid. The curve traced out by a point on the circumference of a circle which rolls, without slipping, upon a
straight line, is called a cycloid.

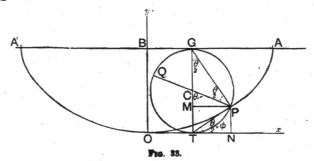

Fig. 33.

Let GPT be any position of the rolling circle, P the genera-
ting point, G the point of contact of the circle with the fixed
straight line ABA', AOA' the cycloid, PCQ, GCT diameters of
the circle.

From the way the curve is generated we see that:

(1) AB = A'B = $\frac{1}{2}$ the circumference of the circle.

(2) BO, which bisects AA' at right angles, is an axis of
symmetry.

(3) The arc GP = AG.

(4) The arc PT = the arc QG = BG = OT.

Also G is the centre of instantaneous rotation of the point P.

∴ PG is normal to the curve at P,

and TP is a tangent   ,,   ,,

The equations of the cycloid.

Take OB as axis of y, and a straight line perpendicular to
it at O, and in the plane of the cycloid, as axis of x. Let the
∠ PCT = θ, and draw PN and PM perpendicular to Ox and GT
respectively.

$$x = ON = OT + TN = BG + TN = \text{arc } QG + PM$$

$$= a\theta + a \sin \theta.$$

$$\therefore \ x = a(\theta + \sin \theta). \quad \dots\dots\dots\dots\dots\dots(1)$$

$$y = PN = MT = CT - CM = a - a \cos \theta.$$

$$\therefore \ y = a(1 - \cos \theta). \quad \dots\dots\dots\dots\dots(2)$$

Equations (1) and (2) determine the position of a point on the cycloid when θ is given.

Tangent and normal. (Using the calculus.)

$$\frac{dx}{d\theta} = a + a \cos \theta, \text{ from (1), and } \frac{dy}{d\theta} = a \sin \theta \text{ from (2)}.$$

$$\therefore \ \frac{dy}{dx} = \frac{\dfrac{dy}{d\theta}}{\dfrac{dx}{d\theta}} = \frac{\sin \theta}{1 + \cos \theta} = \frac{2 \sin \frac{\theta}{2} \cos \frac{\theta}{2}}{2 \cos^2 \frac{\theta}{2}} = \tan \frac{\theta}{2};$$

i.e. the slope of the curve at P

$$= \tan \phi = \tan \frac{\theta}{2}, \quad \text{or} \quad \phi = \frac{\theta}{2}.$$

Also the $\angle$ PTN = the $\angle$ PGT $= \dfrac{\theta}{2}$.

$$\therefore \ \text{PT is the tangent to the curve at P.}$$
$$\therefore \ \text{PG \quad ,, \quad normal \quad ,, \quad ,,}$$

The area of the cycloid.

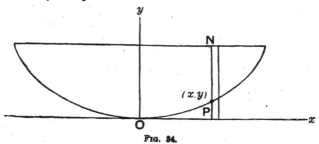

Fig. 34.

The area of the cycloid $= 2 \displaystyle\int_0^{\pi a} PN \, dx = 2 \displaystyle\int_0^{\pi a} (2a - y) \, dx.$

Now $\quad x = a(\theta + \sin \theta) \quad$ and $\quad y = a(1 - \cos \theta).$

$\therefore \ 2a - y = a(1 + \cos \theta) \quad$ and $\quad \dfrac{dx}{d\theta} = a + a \cos \theta.$

Also when $x = \pi a$, $\theta = \pi$, and when $x = 0$, $\theta = 0$.

$$\therefore \text{ the area} = 2\int_0^\pi a^2(1 + \cos\theta)^2 d\theta$$

$$= 2a^2\int_0^\pi \left(1 + 2\cos\theta + \frac{1 + \cos 2\theta}{2}\right)d\theta = 3\pi a^2.$$

The length of the cycloid.   See Example 3, p. 103.

The Logarithmic, or Equiangular, Spiral, whose equation is $r = a^\theta$

In this curve $\tan\psi = \dfrac{r\,d\theta}{dr}$ (p. 102)

$$= \frac{1}{\log_e a} \quad (\text{p. 58}),$$

i.e. the angle between the tangent and radius-vector is constant.

CHAPTER XVI.

SERIES FOR $\sin\theta$, $\cos\theta$, $\tan^{-1}x$. SIMPLE HARMONIC MOTION.

Series for $\sin\theta$ and $\cos\theta$.

Assuming that $\sin\theta$ can be expressed in a series of ascending powers of θ,

let $\sin\theta \equiv a_0 + a_1\theta + \dfrac{a_2\theta^2}{\underline{|2}} + \dfrac{a_3\theta^3}{\underline{|3}} + \dfrac{a_4\theta^4}{\underline{|4}} + \dfrac{a_5\theta^5}{\underline{|5}} + \ldots$ *ad inf.* ...(1)

When $\theta = 0$, $\sin\theta = 0$; $\therefore a_0 = 0$.

Differentiating both sides of equation (1), we have

$$\cos\theta \equiv a_1 + a_2\theta + \dfrac{a_3\theta^2}{\underline{|2}} + \dfrac{a_4\theta^3}{\underline{|3}} + \dfrac{a_5\theta^4}{\underline{|4}} + \ldots \qquad\ldots\ldots\ldots(2)$$

When $\theta = 0$, $\cos\theta = 1$. $\therefore a_1 = 1$.

Differentiating both sides of equation (2), we have

$$-\sin\theta \equiv a_2 + a_3\theta + \dfrac{a_4\theta^2}{\underline{|2}} + \dfrac{a_5\theta^3}{\underline{|3}} + \ldots \qquad\ldots\ldots\ldots(3)$$

$\therefore$ from (1) and (3),

$$a_2 + a_3\theta + \dfrac{a_4\theta^2}{\underline{|2}} + \dfrac{a_5\theta^3}{\underline{|3}} + \ldots \equiv -a_0 - a_1\theta - \dfrac{a_2\theta^2}{\underline{|2}} - \dfrac{a_3\theta^3}{\underline{|3}} - \ldots$$

Equating coefficients in this identity, we have

$$a_2 = -a_0 = 0, \quad a_3 = -a_1 = -1, \quad a_4 = -a_2 = 0,$$
$$a_5 = -a_3 = 1, \text{ and so on.}$$

$$\therefore \sin\theta = \theta - \dfrac{\theta^3}{\underline{|3}} + \dfrac{\theta^5}{\underline{|5}} - \dfrac{\theta^7}{\underline{|7}} + \ldots .$$

Also from equation (2), we have

$$\cos \theta = 1 - \frac{\theta^2}{\underline{|2}} + \frac{\theta^4}{\underline{|4}} - \frac{\theta^6}{\underline{|6}} + \dots .$$

The series for $\cos \theta$ might be found in the same way as that for $\sin \theta$ by assuming that

$$\cos \theta \equiv a_0 + a_1 \theta + a_2 \frac{\theta^2}{\underline{|2}} + \dots .$$

It must be remembered that the above proofs are incomplete, for the series have not been proved to be convergent.

Series for $\tan^{-1} x$.

By the Binomial Theorem, if x^2 is less than unity,

$$\frac{1}{1+x^2} = 1 - x^2 + x^4 - x^6 + x^8 - \dots \textit{ ad inf.}$$

$\therefore$ by integration,

$$\tan^{-1} x = x - \frac{x^3}{3} + \frac{x^5}{5} - \frac{x^7}{7} - \dots \textit{ ad inf.}$$

No constant is necessary, for when $x = 0$, $\tan^{-1} x = 0$.

SIMPLE HARMONIC MOTION.

N.B. $\dfrac{d^2x}{dt^2}$ is accel^n in the positive direction, *i.e.* along Ox.

$\therefore \; -\dfrac{d^2x}{dt^2}$,,   ,, negative   ,,   ,, xO.

$\dfrac{dx}{dt}$ is velocity ,, positive   ,,   ,, Ox.

$\therefore \; -\dfrac{dx}{dt}$,,   ,, negative   ,,   ,, xO.

If a point moves in a straight line so that its acceleration at any time varies as its distance from a fixed point in that line, and is always towards that point, the point is said to move in **Simple Harmonic Motion.**

If O is the fixed point, and P the position of the moving point at time t, let OP $=x$. The acceleration is from P to O,

$$\therefore \frac{d^2x}{dt^2} = -\omega^2 x, \text{ where } \omega \text{ is constant.}$$

$$\frac{dx}{dt}\frac{d^2x}{dt^2} = -\omega^2 x\frac{dx}{dt}.$$

[In connection with this equation it must be remembered that the acceleration changes its direction every time the point passes through the point O.]

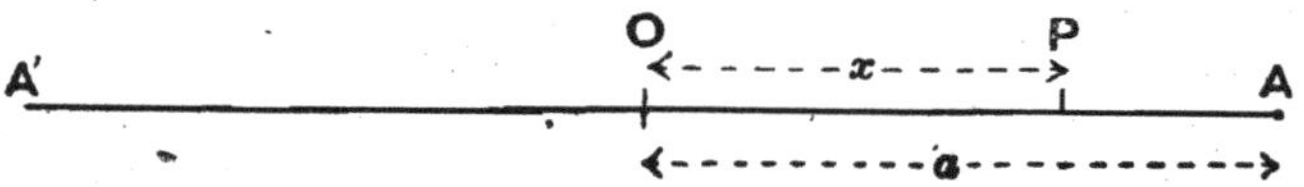

FIG. 25.

Integrating, we have $\dfrac{1}{2}\left(\dfrac{dx}{dt}\right)^2 = -\dfrac{\omega^2 x^2}{2} + C$, where C is constant.

$$\left[\text{Or,} \qquad\qquad v\frac{dv}{dx} = -\omega^2 x. \right.$$

$$\text{Integrating, we have} \qquad \frac{v^2}{2} = -\omega^2 \frac{x^2}{2} + C. \left.\right]$$

$$i.e. \quad \frac{1}{2}\left(\frac{dx}{dt}\right)^2 = -\omega^2 \frac{x^2}{2} + C.$$

If $x = a$ when $v = 0$, $\qquad 0 = -\dfrac{\omega^2 a^2}{2} + C.$

$$\therefore \frac{1}{2}\left(\frac{dx}{dt}\right)^2 = \frac{\omega^2}{2}(a^2 - x^2)$$

and $$\frac{dx}{dt} = \pm \omega\sqrt{a^2 - x^2}$$

If x is positive, v is in the direction PO; therefore we must take the negative sign.

$$\therefore \frac{dx}{dt} = -\omega\sqrt{a^2 - x^2};$$

$$i.e. \quad v = -\omega\sqrt{a^2 - x^2}.$$

$$\therefore -\frac{dx}{\sqrt{a^2 - x^2}} = \omega\,dt.$$

Integrating, we have $\cos^{-1}\left(\dfrac{x}{a}\right) = \omega t + k$.

If $t = 0$ when $x = a$, $k = 0$

$$\therefore \cos^{-1}\left(\dfrac{x}{a}\right) = \omega t \quad \text{or} \quad x = a \cos(\omega t).$$

In the figure, let $OA = OA' = a$.

Let us examine the equations

$$x = a \cos(\omega t), \quad \dots\dots\dots\dots\dots\dots\dots\dots(1)$$
$$v = -\omega\sqrt{a^2 - x^2}. \quad \dots\dots\dots\dots\dots\dots(2)$$

We see that x cannot be greater than a, or less than $-a$.

When $t = 0$, $x = a$ from (1), and $v = 0$ from (2).

From $t = 0$ to $t = \dfrac{\pi}{2\omega}$,

x decreases from a to 0,

and v increases numerically from 0 to ωa, *but is negative,*

i.e. in the direction AO.

From $t = \dfrac{\pi}{2\omega}$ to $t = \dfrac{\pi}{\omega}$,

x changes from 0 to $-a$,

and v decreases numerically from ωa to 0, *but is negative.*

i.e. in the direction OA'.

At the point A' the velocity is zero, and the acceleration is from A' to O; thus we see that the motion is now reversed, *i.e.* the motion from A' to A is similar to that from A to A'.

The point reaches the point A again after a time $\dfrac{2\pi}{\omega}$, and after that, the whole motion from rest (at A) to rest (at A') and back again is repeated.

The time from rest to rest and back again is called the *periodic time.*

EXAMPLES XVI.

1. A particle is moving in a straight line, and at time t its distance x from a fixed point O in the line is given by $x = a \cos(nt + m)$: find its velocity v and its acceleration f at any time.

2. Show that each of the equations (i) $x = A \cos(nt + B)$; (ii) $x = C \sin(nt + D)$; (iii) $x = E \cos nt + F \sin nt$, gives rise to the equation $\dfrac{d^2x}{dt^2} + n^2x = 0$, where A, B, C, D, E, F are constants.

3. A particle moves in a straight line, and when it is at a distance of x ft. from a fixed point in that line its acceleration is $4x$ ft.-sec. units towards that point; and when it is 3 ft. from that point its velocity is zero. Find its velocity when it is (i) at a distance x ft. from that point; (ii) at a distance 2 ft. from that point; (iii) at a distance -1 ft. from that point; (iv) and at the fixed point; (v) find its periodic time.

4. A T-shaped piece has a slot cut in it along AB (see Fig. 36). The upright CD slides vertically between guides. A pin P, which turns about a fixed centre O

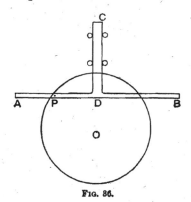

Fig. 36.

with uniform speed, slides in the slot. Determine the motion of C as OP revolves, and find the vertical distance moved through by C during a half-revolution.

5. A point P revolves with uniform speed (u) in a circle, and PQ is drawn perpendicular to a fixed diameter AB. Prove that Q moves along AB in Simple Harmonic Motion, and determine its periodic time.

CHAPTER XVII.

MISCELLANEOUS INTEGRALS AND INTEGRALS OF THE FORM $\int \dfrac{dx}{\pm x^3 \pm a^3}$, $\int \dfrac{dx}{\sqrt{\pm a^2 \pm x^2}}$, $\int \sqrt{\pm a^2 \pm x^2}\, dx$.

IF the integral can be arranged in the form:

(1) $\int \dfrac{dx}{x^2 \sim a^2}$, use partial fractions. See p. **71.**

(2) $\int \dfrac{dx}{x^2 + a^2}$, the integral is $\dfrac{1}{a} \tan^{-1}\dfrac{x}{a}$.

(3) $\int \dfrac{dx}{\sqrt{a^2 - x^2}}$, the integral is $\sin^{-1}\dfrac{x}{a}$.

(4) $\int \dfrac{dx}{\sqrt{x^2 \pm a^2}}$, let $u - x = \sqrt{x^2 \pm a^2}$. See p. **71.**

(5) $\int \sqrt{a^2 - x^2}\, dx$, let $x = a \sin \theta$. See p. 70.

(6) $\int \sqrt{x^2 \pm a^2}\, dx$, use integration by parts. See p. 86.

In the following examples, some of the working is left to the student.

To find the value of $\displaystyle\int \dfrac{dx}{4x^2 + 5x + 6}$.

$$4x^2 + 5x + 6 = 4\left(x^2 + \frac{5x}{4} + \frac{3}{2}\right) = 4\left[\left(x + \frac{5}{8}\right)^2 + \frac{3}{2} - \frac{25}{64}\right]$$

$$= 4\left[\left(x + \frac{5}{8}\right)^2 + \frac{71}{64}\right]. \quad \left(\text{Now let } x + \frac{5}{8} = u.\right)$$

The given integral $= \dfrac{2}{\sqrt{71}} \tan^{-1} \dfrac{8x+5}{\sqrt{71}}$.

To find the value of $\displaystyle\int \dfrac{dx}{2x^2 + 4x + 1}$.

The given integral $= \dfrac{1}{2}\displaystyle\int \dfrac{dx}{x^2 + 2x + \frac{1}{2}} = \dfrac{1}{2}\displaystyle\int \dfrac{dx}{(x+1)^2 - \frac{1}{2}}$.

(Now let $x + 1 = u$.)

The reqd. integral $= \dfrac{1}{2\sqrt{2}} \log \left(\dfrac{x + 1 - \dfrac{1}{\sqrt{2}}}{x + 1 + \dfrac{1}{\sqrt{2}}} \right)$.

To find the value of $\displaystyle\int \dfrac{dx}{\sqrt{3x - 2x^2 + 1}}$.

The given integral $= \dfrac{1}{\sqrt{2}}\displaystyle\int \dfrac{dx}{\sqrt{\dfrac{3x}{2} - x^2 + \dfrac{1}{2}}}$

$$= \dfrac{1}{\sqrt{2}}\displaystyle\int \dfrac{dx}{\sqrt{\dfrac{17}{16} - \left(x - \dfrac{3}{4}\right)^2}}.$$

$\left(\text{Now let } x - \dfrac{3}{4} = u.\right)$

The reqd. integral $= \dfrac{1}{\sqrt{2}} \sin^{-1}\left(\dfrac{4x - 3}{\sqrt{17}}\right)$.

To find the value of $\displaystyle\int \dfrac{2\,dx}{\sqrt{2x^2 + 4x + 3}}$.

The given integral $= \displaystyle\int \dfrac{\sqrt{2}\,dx}{\sqrt{x^2 + 2x + \frac{3}{2}}} = \sqrt{2}\displaystyle\int \dfrac{dx}{\sqrt{(x+1)^2 + \frac{1}{2}}}$.

(Now let $x + 1 = u$.)

The reqd. integral $= \sqrt{2} \log \left[x + 1 + \sqrt{x^2 + 2x + \tfrac{3}{2}} \right]$.

$$\int \sec x \, dx = \int \frac{\cos x}{1 - \sin^2 x} \, dx$$

$$= \int \frac{du}{1 - u^2} \quad \text{if } \sin x = u$$

$$= \frac{1}{2} \log \left(\frac{1+u}{1-u} \right) = \frac{1}{2} \log \left(\frac{1+\sin x}{1-\sin x} \right) = \frac{1}{2} \log \left(\frac{\overline{1+\sin x}|^2}{\cos^2 x} \right)$$

$$= \log \left(\frac{1+\sin x}{\cos x} \right) = \log (\sec x + \tan x).$$

Second Method.

$$\int \sec x \, dx = \int \frac{dx}{\cos^2 \frac{x}{2} - \sin^2 \frac{x}{2}} = \int \frac{\sec^2 \frac{x}{2} \, dx}{1 - \tan^2 \frac{x}{2}} = 2 \int \frac{du}{1 - u^2} \quad \text{if } \tan \frac{x}{2} = u$$

$$= \log \left(\frac{1 + \tan \frac{x}{2}}{1 - \tan \frac{x}{2}} \right) = \frac{1}{2} \log \left(\frac{\cos \frac{x}{2} + \sin \frac{x}{2}}{\cos \frac{x}{2} - \sin \frac{x}{2}} \right)^2 = \frac{1}{2} \log \left(\frac{1 + \sin x}{1 - \sin x} \right),$$

as before

Find the value of $\displaystyle \int \sqrt{3x^2 + 6x + 13} \, dx.$

The integral $= \sqrt{3} \displaystyle\int \sqrt{x^2 + 2x + \tfrac{13}{3}} \, dx = \sqrt{3} \displaystyle\int \sqrt{(x+1)^2 + \tfrac{10}{3}} \, dx$

(Now, if we take $x + 1 = u$, the integral is in the form

$$\sqrt{3} \int \sqrt{x^2 + a^2} \, dx.)$$

$$= \sqrt{3} \left[\frac{x+1}{2} \sqrt{x^2 + 2x + \tfrac{13}{3}} + \tfrac{5}{3} \log \left(x + 1 + \sqrt{(x+1)^2 + \tfrac{10}{3}} \right) \right].$$

Find the value of $\displaystyle \int \sqrt{1 - 4x - 2x^2} \, dx.$

The integral $= \sqrt{2} \displaystyle\int \sqrt{\tfrac{3}{2} - (x+1)^2} \, dx$

(If we take $x+1=u$, the integral is in the form

$$\sqrt{2}\int \sqrt{a^2-x^2}\,dx.)$$

$$=\sqrt{2}\left[\frac{x+1}{2}\sqrt{\tfrac{3}{2}-(x+1)^2}+\tfrac{3}{4}\sin^{-1}\left(\frac{x+1}{\sqrt{\tfrac{3}{2}}}\right)\right].$$

Find the value of $\displaystyle\int \frac{d\theta}{8-5\sin 2\theta}.$

The integral $\displaystyle=\int\frac{d\theta}{8\cos^2\theta+8\sin^2\theta-10\sin\theta\cos\theta}$

$$=\frac{1}{8}\int\frac{\sec^2\theta\,d\theta}{\tan^2\theta-\tfrac{5}{4}\tan\theta+1}$$

$$=\frac{1}{8}\int\frac{\sec^2\theta\,.\,d\theta}{(\tan\theta-\tfrac{5}{8})^2+\tfrac{39}{64}}.$$

Now let $\tan\theta-\tfrac{5}{8}=u$, so that $\sec^2\theta\,.\,d\theta=du$.
The integral

$$=\frac{1}{8}\int\frac{du}{u^2+\tfrac{39}{64}}=\frac{1}{8}\cdot\frac{8}{\sqrt{39}}\tan^{-1}\left(\frac{8u}{\sqrt{39}}\right)\left[\int\frac{dx}{x^2+a^2}=\frac{1}{a}\tan^{-1}\left(\frac{x}{a}\right)\right]$$

$$=\frac{1}{\sqrt{39}}\tan^{-1}\left[\frac{8\tan\theta-5}{\sqrt{39}}\right].$$

EXAMPLES XVII.

1. $\displaystyle\int\sin^3x\,dx.$ 2. $\displaystyle\int\sin x\cos 2x\,dx.$ 3. $\displaystyle\int\sin^3x\cos^2x\,dx.$

4. $\displaystyle\int\frac{2-x}{3-x^2}\,dx.$ 5 $\displaystyle\int\frac{d\theta}{\sin 3\theta}.$ 6. $\displaystyle\int\frac{\tan x\,.\,dx}{1+\cos^2x}.$

7. $\displaystyle\int\frac{x\,dx}{(2x-3)(x^2+2)}.$ 8. $\displaystyle\int\tan(x+a)\,dx.$

9. $\displaystyle\int\cos 2x\sin^2x\,dx.$ 10. $\displaystyle\int\frac{\cos^3x}{\sqrt{\sin x}}\,dx.$

11. $\displaystyle\int\frac{d\theta}{9\cos^2\theta+16\sin^2\theta}.$ 12. $\displaystyle\int\frac{dx}{(1-x)\sqrt{1-x^2}}.$ (Let $x=\cos\theta.$)

B.J. K

Integrate the following expressions:

13. $\dfrac{1}{x^2 - 7}$.

14. $\dfrac{2}{\sqrt{7 + 12x - 4x^2}}$.

15. $\dfrac{1}{\sqrt{x^2 + 4x + 5}}$.

16. $\sqrt{x^2 + 9}$.

17. $\dfrac{1}{\sqrt{4x^2 + 1}}$.

18. $\sqrt{1 - 4x^2}$.

19. $\dfrac{3}{2x^2 - 2x + 5}$.

20. $\dfrac{1}{\sqrt{6x - 9x^2}}$.

21. $\dfrac{1}{4x^2 - 9}$.

22. $\sec 2x$.

23. $\dfrac{1}{3 - \sin 2x}$.

24. $\dfrac{1}{5 + 3 \cos x}$.

25. $\dfrac{1}{\sqrt{4x - x^2}}$.

26. $\dfrac{1}{4x - x^2}$.

27. $\dfrac{1}{x^2 + x + 5}$.

28. $\dfrac{1}{\sqrt{x^2 + 2x + 2}}$.

29. $\dfrac{\log_e x}{x}$.

30. $\dfrac{\log_a x}{x}$.

31. $\dfrac{1}{(x^2 + 9)^2}$.

32. $\dfrac{1}{(x^2 - 1)^2}$. ($x > 1$. Use partial fractions.)

33. $\dfrac{3x^3}{x - 2}$.

34. $5x^3 \sqrt{9 - x^2}$.

35. $\dfrac{1}{9 + 8 \cos x}$.

36. $\dfrac{1}{8 + 9 \cos x}$.

37. $\dfrac{1}{x^2 \sqrt{4 - x^2}}$.

38. $\sec \dfrac{x}{2}$.

39. $\sec^3 x$.

40. $\dfrac{1}{x \sqrt{4 - x^2}}$.

REVISION PAPERS.

REVISION PAPER I.

1. Find the differential coefficients with respect to x of

$$\frac{x^2 - a^2}{x^2 + a^2}, \quad e^{-x}\cos(a - 4x), \quad \log\tan\left(\frac{\pi}{4} + \frac{x}{2}\right).$$

2. The space-time equation of a moving particle is

$$x = 5 + 6t + 4t^2$$

in ft. sec. units; find (1) the velocity and acceleration of the particle when $t = 3$; (2) the space described and the velocity added during the next second.

Explain why one pair of values in (1) and (2) agree.

3. Prove that the equation of the tangent at the point (x_1, y_1) on the curve $y = x^3 - 3x$ may be written

$$y - 3(x_1^2 - 1)x + 2x_1^3 = 0.$$

Write down the equation of the normal at the same point.

4. Obtain the following integrals :

$$\int \frac{dx}{\sqrt{1 - 4x}}, \quad \int \frac{1 - x}{\sqrt{4 - x^2}}\,dx, \quad \int \frac{dx}{4 + 7x^2}.$$

5. Find the volume, to the nearest c. cm., of the greatest right circular cone which can be cut from a solid sphere of radius 10 cm.

REVISION PAPER II.

1. Find the slope of the curve $y = 8x^2 - 1$ at the point where $x = 1.5$. Also find the average slope from $x = 1$ to $x = 2$, and explain why the two results are the same.

2. If $y = (1 - 2x)^{\frac{1}{2}}$, find $\dfrac{dy}{dx}$ from first principles.

Also find the differential coefficients of

$$(1)\ \frac{a+x}{a-x}, \qquad (2)\ \frac{1}{\sqrt{x+a}+\sqrt{x-a}}.$$

3. A curve passes through the point $(1, 1)$, and its slope at any point (x, y) is $3x - 5$; find the equation of the curve.

4. Show in a diagram the area represented by

$$\int_{-5}^{12} \sqrt{169 - x^2}\, dx,$$

and find this area (to the nearest integer) in any way you please.

5. Find the values of the following integrals:

$$(1)\ \int \frac{dx}{x^2 - 4x - 21}, \qquad (2)\ \int \frac{x\,dx}{1 + 5x^2}, \qquad (3)\ \int \frac{dx}{1 + 5x^2}.$$

REVISION PAPER III.

1. If $y = \dfrac{1}{x-2}$, write down the values of

$$\frac{dy}{dx}, \quad \frac{d^2y}{dx^2}, \quad \frac{d^3y}{dx^3}, \quad \text{and} \quad \frac{d^ny}{dx^n}.$$

2. ABC is a triangle, B being an obtuse angle, and CN is drawn perpendicular to AB produced. If AC and BC are stretched by small amounts db and da, find the corresponding increase of BN.

3. Find the minimum value of $x^2 + 16 + \dfrac{4}{x^2}$, (1) by algebra, (2) by using the calculus. Has it any maximum value?

4. Find the values of the following integrals:

$$(1)\ \int \sin^2 3x\, dx, \qquad (2)\ \int \frac{2x-3}{\sqrt{1-x^2}}\, dx, \qquad (3)\ \int \frac{dx}{\sqrt{16 + x^2}}.$$

5. Find the area between the curve $y^2 = 16x$, the axis of x, and the chord joining the two points $(4, 8)$, $(1, -4)$.

REVISION PAPER IV.

1. If $y = \dfrac{1}{ax+b}$, find the values of $\dfrac{dy}{dx}$, $\dfrac{d^2y}{dx^2}$, $\dfrac{d^3y}{dx^3}$, and $\dfrac{d^ny}{dx^n}$.

2. Show from graphical considerations that

$$\int_a^b f(x)\,dx = \int_0^{b-a} f(x+a)\,dx.$$

3. In the curve $y = e^{-ax}\sin bx$, prove that maximum and minimum values of y are obtained from values of x given by

$$\tan bx = \frac{b}{a}.$$

4. Two curves, $yx^2 = $ a constant and $yx = $ a constant, pass through the point P(3, 4); find the angle at which they intersect at P.

Find also the length intercepted on the axis of x between the two normals drawn to the curves at P.

5. Find the values of the following:

(1) $\displaystyle\int \frac{\sqrt{x}}{4+x}\,dx.$ (Let $x = u^2$.) (2) $\displaystyle\int \frac{\sqrt{x}+1}{\sqrt{x}-1}\,dx.$

(3) $\displaystyle\int \frac{x}{\sqrt{x-1}}\,dx.$

REVISION PAPER V.

1. We know that $\sin x$ increases as x varies from 0 to $\dfrac{\pi}{2}$.

Show from the principles of the Differential Calculus that the rate of increase diminishes. Verify this result graphically.

2. Differentiate $\dfrac{1}{x^2-5x+6}$ with respect to x, (1) by treating it as a power of x^2-5x+6, (2) by first resolving it into partial fractions.

3. Show from graphical considerations that

$$\int_a^b f(nx)\,dx = \frac{1}{n}\int_{na}^{nb} f(x)\,dx.$$

Also prove the same by another method.

4. Find the values of the following : -

(1) $\int e^x (\cos x + \sin x)\, dx$,   (2) $\int \dfrac{x\, dx}{(a + bx)^{\frac{1}{2}}}$,   (3) $\int \dfrac{dx}{4 + 3 \cos x}$.

5. Find the area enclosed by the two parabolas

$$y^2 = -4(x-1), \quad y^2 = 4(x+1).$$

REVISION PAPER VI.

1. In the same diagram, draw accurate graphs of the curves $y = x^3$, $y^2 = x^3$.

2. If $y = a \sin (x + b)$, prove that $\dfrac{d^2 y}{dx^2} + y = 0$.

Also if $y = \sin(bx + c)$, prove that $\dfrac{d^2 y}{dx^2}(1 - y^2) + y\left(\dfrac{dy}{dx}\right)^2 = 0$.

3. Use " integration by parts " to find the value of

$$\int x \sqrt{x+1}\, dx.$$

4. Throughout a circular area of radius a each element of area is multiplied by the square of its distance from the centre : prove that the sum of these products is $\dfrac{\pi a^4}{2}$.

5. Prove that the area enclosed by the curves $y^2 = 4ax$, $x^2 + y^2 = 5a^2$ is equal to $\dfrac{2a^2}{3} - 5a^2 \tan^{-1} 2$.

REVISION PAPER VII.

1. By differentiation eliminate the constant from the equation $y = \sqrt{a^2 + x^2}$.

Solve the differential equation $y\dfrac{dy}{dx} + x = 0$, i.e. find y in terms of x and a constant.

2. Water is poured into a conical vessel whose axis is vertical at the rate of 10 c. ft. per minute. If 30° is the semi-vertical angle of the cone, find the rate at which the depth is increasing when that depth is x feet.

· 3. Draw a sketch of the curve $y = 3\left[1 + \cos\left(x + \dfrac{\pi}{4}\right)\right]$ between the ordinates $x = 0$ and $x = \dfrac{\pi}{2}$. Find the area bounded by the curve, these ordinates, and the axis of x.

4. Find the maximum and minimum values of
$$2x^2(x^2 - 8x - 54) + 400.$$

5. Obtain the following integrals:

(1) $\displaystyle\int \frac{\cos x}{1 + 4\sin^2 x}\, dx,$　(2) $\displaystyle\int \frac{3 + 5x}{1 + 4x^2}\, dx,$　(3) $\displaystyle\int e^{2x} \sin x\, dx.$

REVISION PAPER VIII.

1. Find the differential coefficient of
$$(1)\ \sin(x + a),\qquad (2)\ (1 - x^2)^{\frac{2}{3}},$$
from first principles.

2. Find the points on the curve $y = x^3 - 9x^2 + 15x + 12$, where y is a maximum or minimum.

3. Prove by the Calculus that $x - \log_e(1 + x)$ is positive for all positive values of x.

4. Find the area contained by the curve
$$y = (x - 1)(x - 2)(x - 3),$$
and the axes of co-ordinates.

5. If V c. ft. is the volume of a solid and x ft. its height, it is known that $V = 16\pi x^2$, what is the meaning of $\dfrac{dV}{dx}$? Find the area of a section at a height of 4 inches.

REVISION PAPER IX.

1. Find the differential coefficients of
$$(1)\ \log\frac{a + bx}{a - bx},\qquad (2)\ e^{x \log x}$$

2. Find the value of $\int \sin x \cos x\, dx$, (1) by taking $u = \sin x$, (2) by taking it as $\int \frac{\sin 2x}{2}\, dx$, and account for the different results.

3. Find the values of

(1) $\int \frac{dx}{\sqrt{2ax - x^2}}$, (2) $\int \frac{dx}{x\sqrt{2ax - a^2}}$. $\left(\text{Take } u = \frac{x-a}{x}.\right)$

4. Prove that the length of the portion of a tangent to the curve $x^{\frac{2}{3}} + y^{\frac{2}{3}} = a^{\frac{2}{3}}$ intercepted between the axes of co-ordinates is constant.

5. Find the points on the curve $y = x^3 - 3x^2$, where the tangent is parallel to the axis of x, and draw the graph of the curve.

REVISION PAPER X.

1. Find the differential coefficients of

(1) $\log (2x + \sqrt{1 + 4x^2})$,   (2) $\sqrt{x + \frac{1}{x}}$.

2. Prove that

$$\int \frac{x\, dx}{\sqrt{1 + x^4}} = \frac{1}{2} \int \frac{du}{\sqrt{1 + u^2}},$$

and hence find its value.

3. A body of mass m, starting with velocity u, falls freely under the action of gravity. Find the rate at which its kinetic energy is increasing at any point, (1) per unit of space, (2) per unit of time.

4. Draw the curve $y^2 = x^2(3 - x)$, and determine the slopes of the tangents at the origin.

5. If s is the length of an arc of the parabola $y^2 = 4ax$, prove that

$$\frac{ds}{dx} = \sqrt{\frac{a + x}{x}}.$$

REVISION PAPER XI.

1. Find the differential coefficients of

$$(1)\ (ax)^{bx}, \qquad (2)\ \tan^{-1}\frac{4x}{1-4x^2}.$$

2. Find the values of $(1)\ \displaystyle\int \frac{x^2\,dx}{1+4x}, \qquad (2)\ \displaystyle\int_{-2}^{10} x\sqrt{6+x}\,dx.$

3. If a body starts from rest under an acceleration which increases uniformly by 3 ft. per sec. per sec., find its velocity after 4 secs. if its initial acceleration is 5 ft. per sec. per sec.

4. If the radius (r) of a circle increases, find the rate (with respect to r) at which its area (A) increases. Prove that if, at any stage, r is increased by 3 per cent. of its value, A is increased by 6 per cent. of its value.

5. In the ellipse $\dfrac{x^2}{a^2}+\dfrac{y^2}{b^2}=1$, if $x=a\cos\phi$, prove that

$$\frac{ds}{d\phi}=a\sqrt{1-e^2\cos^2\phi}.$$

REVISION PAPER XII.

1. Prove that $\displaystyle\int_{1}^{3}\frac{dx}{1+x^2}=0\cdot46$ approx.,

and that $\displaystyle\int_{a}^{b} f(x)\,dx=\int_{a+b}^{2a} f(x-a)\,dx.$

2. A bucket (without a lid), of cylindrical shape, is made from 5 sq. feet of metal. Find its maximum volume to the nearest tenth of a cubic foot.

3. If $\dfrac{dy}{dx}+\dfrac{ax}{y}=0$, prove that the value of ax^2+y^2 is constant.

4. Find the area contained by the curve $y=\dfrac{5}{x^2-1}$, the axis of x, and the lines $x=2$ and $x=5$.

5. A rectangle 2 ft. by 4 ft. is immersed vertically in water with a shorter side in the surface. Find (1) the total pressure upon it, (2) the depth of its centre of pressure.

REVISION PAPER XIII.

1. Find the values of

$$(1) \int_0^1 \frac{(8x - 3)\,dx}{(4x^2 - 3x - 5)^2}, \quad (2) \int_{-\frac{\pi}{2}}^{\frac{\pi}{2}} \cos^2 4\theta\,d\theta.$$

2. One end of a dock, 30 ft. deep, is in the form of a trapezium, whose upper side is 40 ft. and lower side 30 ft. long. Find the resultant pressure on this end of the dock when it is full of water.

3. Find the maximum area contained by the curve

$$4y = 4x - x^2,$$

the axis of x and two ordinates one unit apart.

4. If $\dfrac{d^2y}{dx^2} = \sin 2x$, prove that $y = ax + b - \dfrac{\sin 2x}{4}$, where a and b are constants.

5. A rectangle 1 ft. by 2 ft. is immersed vertically in water with its shorter upper side at a depth of 5 ft. Find (1) the total pressure upon it, (2) the depth of its centre of pressure.

REVISION PAPER XIV.

1. Find the values of

$$(1) \int e^x (2 \cos 2x + \sin 2x)\,dx.$$

$$(2) \int \sec^3 \theta \operatorname{cosec} \theta\,d\theta. \quad \text{(Take } u = \tan \theta.)$$

2. A straight line AB makes with the axes of co-ordinates a triangle AOB whose area is constant. If the point A moves along Ox, prove that the ratio of its speed to the speed of B along yO is $\dfrac{OA}{OB}$.

3. An error of 0·3 per cent. is made in measuring the radius of a sphere; find the approximate percentage error in its calculated volume.

4. A mass of m lb. lying on a rough horizontal plane, whose coefficient of friction is μ, is pulled slowly along the plane by a rope which passes over a pulley 6 ft. above the ground. If x ft. is the distance of the mass from the vertical through the pulley, prove that the work done in moving the mass through a distance a is

$$\int_{b-a}^{b} \frac{m\mu gx}{x+6\mu}\,dx,$$

if the mass starts from the point where $x = b$.

Find the work done when $m = 60$, $\mu = \frac{1}{3}$, $a = 4$, and $b = 12$.

5. Temperature being measured in degrees Centigrade, the rate of cooling of a body varies as the excess of its temperature above that of the room in which it is placed. Express this as a differential equation, and if a is the temperature of the room, prove that

$$T - a = be^{-kt},$$

where b and k are constants, T is the temperature of the body, and t the time of cooling.

REVISION PAPER XV.

1. Find the values of

$$(1)\ \int \frac{dx}{(x^2+9)(x^2+16)}, \qquad (2)\ \int \frac{x^2\,dx}{\sqrt{1-x^2}}.$$

2. If the edge of a cube increases at the rate of 0·002 in. per sec., find the rate at which its volume is increasing when the edge is 18 inches.

3. A circular sector has a perimeter of 4 feet : find its maximum area.

4. A chain weighing 20 lb. per ft. length is coiled up at the foot of a rough inclined plane of elevation 10°, and whose coefficient of friction is 0·5. Find the work done in pulling the chain slowly up the plane by one end until there are 40 ft. of chain on the plane.

5. OB is the diameter of a semi-circle OBC, of radius a, and centre A. If the quadrant ABC rotates about an axis through O and parallel to AC, prove that the volume generated is equal to $\pi a^3 \left(\dfrac{2}{3} + \dfrac{\pi}{2} \right)$.

INDEX.

ANSWERS.

Examples I. a. (p. 9.)

1. $-a$. 2. 6. 3. 2. 4. 2.

5. $2ax$. 6. $3x^2$. 7. $2ax + b$. 8. $7x^6$.

9. $54x^5$. 10. $3ax^2 + 2bx$. 11. $8 - 28t$. 12. $u + ft$.

13. $-\dfrac{1}{x^2}$. 14. $-\dfrac{2}{x^3}$. 15. $2x + 5$. 16. $-\dfrac{1}{(x+2)^2}$.

17. $-\dfrac{4}{(x-2)^2}$. 18. $\dfrac{1}{(1-x)^2}$. 19. $-\dfrac{20}{v^3}$. 20. $-\dfrac{8}{(v-2)^2}$.

21. $-\dfrac{3}{(3x-4)^2}$. 22. $6x - 4$. 23. $2x + 3$. 24. $-1 - 4x$.

Examples I. b. (p. 12.)

1. $\dfrac{1}{2\sqrt{x}}$. 2. $\dfrac{5}{2}x^{\frac{3}{2}}$. 3. $-4x^{-5}$. 4. $-nx^{-n-1}$.

5. $18(3x+1)^5$. 6. $\dfrac{4}{(1-x)^5}$. 7. $-\dfrac{5}{(a+x)^6}$. 8. $n(x+a)^{n-1}$.

9. $an(ax+b)^{n-1}$. 10. $a - \dfrac{c}{x^2}$. 11. $-\dfrac{2x}{(1+x^2)^2}$. 12. $\dfrac{-2bx}{(a+bx^2)^2}$.

13. $\dfrac{3}{2\sqrt{3x-5}}$. 14. $\dfrac{x}{\sqrt{1+x^2}}$. 15. $\dfrac{1}{3}(1+x)^{-\frac{2}{3}}$. 16. $\dfrac{5}{3}(2+5x)^{-\frac{2}{3}}$.

17. $-40(5x-7)^{-3}$. 18. $bn(a-bx)^{-n-1}$. 19. $-\dfrac{3}{2}(1-x)^{\frac{1}{2}}$.

20. $(2-3x)^{-\frac{4}{3}}$. 21. $-\dfrac{a}{x^2}$. 22. $-\dfrac{1}{2}ax^{-\frac{3}{2}} + \dfrac{1}{2}bx^{-\frac{1}{2}}$.

23. $\dfrac{1}{2\sqrt{1+x}}$. 24. 4. 25. 3. 26. b. 27. 0.

28. $-\dfrac{x}{y}$. 29. $\dfrac{y}{2x}$. 30. $\dfrac{2ax+b}{2y}$. 31. $\dfrac{b^2x}{a^2y}$.

Examples I. c.   (p. 13.)

1. $-\dfrac{1}{x^3}$.   **2.** $-\dfrac{2}{x^3}$.   **3.** $-4x^{-5}$.   **4.** $7ax^6$.

5. $-7x^{-8}$.   **6.** $\dfrac{5}{2}x^{\frac{3}{2}}$.   **7.** $6x^5 + 5ax^4 - 4bx^{-}$.   **8.** $-anx^{-n-1}$.

9. $2a^2x + b$.   **10.** $-\dfrac{c}{x^2}$.   **11.** $-\dfrac{3a}{x^4}$.   **12.** $2x + a - \dfrac{b}{x^2}$.

13. $-\dfrac{2a^2}{x^3} - \dfrac{b}{x^3}$.   **14.** $(n-1)px^{n-2} + (n-2)qx^{n-3} + (n-3)rx^{n-4}$.

15. $1 + x + \dfrac{x^2}{\underline{2}} + \dfrac{x^3}{\underline{3}}$.   **16.** $\dfrac{x^8}{\underline{9}} - \dfrac{x^8}{\underline{8}} + \dfrac{x^7}{\underline{7}}$.   **17.** $\dfrac{x^{n-1}}{\underline{n-1}} + \dfrac{x^{n-2}}{\underline{n-2}} + \dfrac{x^{n-3}}{\underline{n-3}}$.

18. $-3x^{-4} - x^{-2}$.   **19.** $3 + 4x$.   **20.** $2x$.   **21.** $-\dfrac{a}{x^2}$.   **22.** $2 + 2x$.

23. $-\dfrac{1}{x^3}$.   **24.** $\dfrac{1}{2}x^{-\frac{1}{2}} - \dfrac{1}{2}bx^{-\frac{3}{2}}$.   **25.** $t^2 + at + b$.   **26.** $\dfrac{1}{7\sqrt{x}}$.

27. $-\dfrac{2}{x^3} - \dfrac{1}{x^2}$.   **28.** $-\dfrac{1}{x^2} + 1$.   **29.** $\dfrac{1}{2}x^{-\frac{1}{2}} - \dfrac{1}{2}x^{-\frac{3}{2}}$.   **30.** $\dfrac{1}{3}x^{-\frac{2}{3}} - \dfrac{1}{3}x^{-\frac{4}{3}}$.

31. $1 + \dfrac{1}{x^2}$.   **32.** $2x + \dfrac{1}{x^2}$.   **33.** $8t - 4$.   **34.** $b - \dfrac{c}{t^2}$.

Examples I. d.   (p. 14.)

1. $2a$.   **2.** x.   **3.** $6(x-1)$.

4. $(n-1)x^{n-2} - 6(n-2)x^{n-3} + 7(n-3)x^{n-4}$.   **5.** $x^{n-2} + x^{n-3} + x^{n-4}$.

6. 1.   **7.** $\dfrac{x^3}{\underline{3}} - \dfrac{x^4}{\underline{4}} + \dfrac{x^5}{\underline{5}} - \dfrac{x^6}{\underline{6}}$.   **8.** $-6x^{-4} - 24x^{-5} - 60x^{-6}$.   **9.** f

Examples II. a.   (p. 17.)

1. (a) $\dfrac{2a}{y}$.   (b) $y - y_1 = \dfrac{2a}{y_1}(x - x_1)$, or $yy_1 = 2a(x + x_1)$.

2. (a) $\dfrac{x}{2a}$.   (b) $y - y_1 = \dfrac{x_1}{2a}(x - x_1)$, or $xx_1 = 2a(y + y_1)$.

3. (a) $-\dfrac{x}{4y}$.   (b) $y - y_1 = -\dfrac{x_1}{4y_1}(x - x_1)$, or $xx_1 + 4yy_1 = 16$.

4. (a) $-\dfrac{ax}{by}$.   (b) $y - y_1 = -\dfrac{ax_1}{by_1}(x - x_1)$, or $axx_1 + byy_1 = c$.

5. (a) $-\dfrac{b^2x}{a^2y}$.   (b) $y - y_1 = -\dfrac{b^2x_1}{a^2y_1}(x - x_1)$, or $\dfrac{xx_1}{a^2} + \dfrac{yy_1}{b^2} = 1$.

6. (a) $\dfrac{x}{y}$ (b) $y - y_1 = \dfrac{x_1}{y_1}(x - x_1)$, or $xx_1 - yy_1 = a^2$.

7. (a) $\dfrac{a}{y-b}$.   (b) $y - y_1 = \dfrac{a}{y_1 - b}(x - x_1)$, or $yy_1 = a(x + x_1) + b(y + y_1)$.

8. (a) $\dfrac{2-x}{3}$.   (b) $y - y_1 = \dfrac{2 - x_1}{3}(x - x_1)$, or $xx_1 - 2(x + x_1) + 3(y + y_1) = 0$.

9. (a) $\dfrac{1-x}{2+y}$.   (b) $y - y_1 = \dfrac{1 - x_1}{2 + y_1}(x - x_1)$,
 or $xx_1 + yy_1 - (x + x_1) + 2(y + y_1) = 0$.

10. (a) $\dfrac{2-x}{4(1+y)}$.   (b) $y - y_1 = \dfrac{2 - x_1}{4(1+y_1)}(x - x_1)$,
 or $xx_1 + 4yy_1 = 2(x + x_1) - 4(y + y_1)$.

11. $\dfrac{dy}{dx} = -0\cdot3$.   The tangent cuts the curve.   **12.** $\dfrac{dy}{dx} = 0$.

13. $\left(-\frac{1}{4}, \frac{13}{4}\right), \left(\frac{1}{3}, \frac{2}{27}\right)$.

14. The curve is a parabola, and the chords joining the points where $x = 1$ and $x = 3$ is parallel to the tangent where $x = 2$.

Examples II. b. (p. 25.)

1. $(1, -\frac{1}{3})$. A minimum. **2.** $(1, 1)$. A maximum.
3. $(4, -4)$. A minimum. **4.** $\left(\frac{1}{2}, \frac{3}{2}\right)$. A minimum.
5. $(3, 5)$. A maximum. **6.** $\left(-\dfrac{c}{2b}, -\dfrac{c^2}{4ab}\right)$. A minimum.
7. Maximum, 4; minimum, 0. **8.** Maximum, 18; minimum, 17.
9. Maximum, 162; minimum, 37.
10. Maximum, $6\sqrt{3}(=10\cdot39)$; minimum, $-6\sqrt{3}(=-10\cdot39)$.
11. Maximum, 8; no minimum. **12.** Maximum, -8; minimum, 16.
13. Minimum, $5\cdot37$; no maximum. **14.** Maximum, -2; minimum, 2.

Examples II. c.   (p. 27.)

1. $(0, 0)$. **2.** $(0, 5)$. **3.** $\left(\frac{1}{2}, \frac{3}{2}\right)$. **4.** $(\pm1, -5)$. **5.** $-\dfrac{b}{3a}$.

Examples III. a.   (p. 32.)

1. $44x + 25$. **2.** $83 - 70x$. **3.** $\dfrac{1}{(x+1)^2}$. **4.** $\dfrac{x^2 - 2x}{(x-1)^2}$

5. $\dfrac{24}{(3x+4)^2}$. **6.** $\dfrac{6(1-x)}{(4+2x-x^2)^2}$. **7.** $26(8 - 13x)$. **8.** $3x^2 + 12x + 11$.

9. $\dfrac{n}{x^{n+1}}.$

10. $\dfrac{2nx^{n-1}}{(x^n+1)^2}.$

11. $3(x+1)^2.$

12. $-3(1-x)^2.$

13. $11(x-4)^{10}.$

14. $\dfrac{1}{2\sqrt{x+7}}.$

15. $\dfrac{3}{2}(x-a)^{\frac{1}{2}}.$

16. $\dfrac{-2}{(x-a)^3}.$

17. $-\dfrac{1}{2\sqrt{1-x}}.$

18. $-\dfrac{7}{(x-2)^8}.$

19. $4(9+x)^3.$

20. $-\dfrac{7}{2}(3-x)^{\frac{5}{2}}.$

21. $27(3x-4)^8.$

22. $-11(2-x)^{10}.$

23. $\dfrac{1}{\sqrt{1+2x}}.$

24. $-16(1-2x)^7$

25. $\dfrac{1}{\sqrt{2x+7}}.$

26. $-(1-3x)^{-\frac{2}{3}}.$

27. $5(5-4x)^{-\frac{9}{4}}.$

28. $\dfrac{a}{2\sqrt{ax+b}}.$

29. $-(1+2x)^{-\frac{3}{2}}.$

30. $(3-2x)^{-\frac{1}{2}}.$

31. $-\dfrac{a}{2}(ax+b)^{-\frac{3}{2}}.$

32. $\dfrac{x}{\sqrt{1+x^2}}.$

33. $14x(1+x^2)^6.$

34. $-3x(1-x^2)^{\frac{1}{2}}.$

35. $-\dfrac{2x}{\sqrt{5-2x^2}}.$

36. $\dfrac{2ax+b}{2\sqrt{ax^2+bx+c}}.$

37. $-\dfrac{1}{2}(2ax+b)(ax^2+bx+c)^{-\frac{3}{2}}.$

38. $9(2ax+b)(ax^2+bx+c)^8.$

39. $\dfrac{10x}{3}(5x^2-7)^{-\frac{2}{3}}.$

40. $\dfrac{2}{3}(x-1)(x^2-2x+5)^{-\frac{2}{3}}.$

41. $-\dfrac{1}{3}(2x-5)(x^2-5x+8)^{-\frac{4}{3}}.$

42. $\dfrac{2x^2-1}{\sqrt{x^2-1}}.$

43. $\dfrac{1}{2}\left[\dfrac{1}{\sqrt{x+1}}+\dfrac{1}{\sqrt{x}}\right].$

44. $v+x\dfrac{dv}{dx}.$

45. $3ap^2\dfrac{dp}{dx}.$

47. (1) $2y\dfrac{dy}{dx}$;   (2) $y^2+2xy\dfrac{dy}{dx}$;   (3) $2xy^3+2x^2y\dfrac{dy}{dx}.$

48. $\dfrac{2x+5}{7-2y}.$

49. $-\dfrac{y}{x}.$

50. $\dfrac{2}{y}.$

51. $-\dfrac{b^2x}{a^2y}.$

52. $\dfrac{b^2x}{a^2y}.$

53. $-\dfrac{ax+hy}{hx+by}.$

54. $\dfrac{ay-3x^2}{3y^2-ax}.$

55. $\dfrac{3x^2+4x}{2(x+1)^{\frac{3}{2}}}.$

Examples III. b.   (p. 39.)

1. $2\cos 2x.$

2. $-3\sin 3x.$

3. $2\cos(2x-4).$

4. $3\sin(5-3x).$

5. $\dfrac{1}{2}\cos\dfrac{x}{2}.$

6. $-\dfrac{1}{a}\sin\dfrac{x}{a}.$

7. $\dfrac{1}{5}\sec^2\dfrac{x}{5}.$

8. $\dfrac{1}{3}\sec\dfrac{x}{3}\tan\dfrac{x}{3}.$

9. $5\sec^2 5x.$

10. $3\sec^2(3x-5).$

11. $-\dfrac{1}{3}\operatorname{cosec}^2\left(\dfrac{\pi+x}{3}\right).$

12. $3\sec(3x-7)\tan(3x-7).$

13. $4\operatorname{cosec}(9-4x)\cot(9-4x).$

14. $\sin 2x.$

15. $5\sin 10x.$

16. $-\sin 2x.$

17. $\sin(2x+4).$

18. $-3\sin(6x-8).$

19. $9\cos^2(4-3x)\sin(4-3x)$. **20.** $10\tan(5x-6)\sec^2(5x-6)$.

21. $-\dfrac{\pi}{180}\sin x^\circ$. **22.** $\dfrac{\pi}{180}\sec^2(x^\circ+45^\circ)$. **23.** $-2\cos x(1-\sin x)$.

24. $3\sin x(1-\cos x)^2$. **25.** $-\dfrac{\sec^2 x}{(1+\tan x)^2}$. **26.** $3\sin^2 x\cos x$.

27. $6\sin^2 2x\cos 2x$. **28.** $5\cos 10x+3\cos 6x$. **29.** $3\sin 6x-\sin 2x$.

30. $-\dfrac{3\sin 3x}{2\sqrt{\cos 3x}}$. **31.** $\cos 3x(\sin 3x)^{-\frac{2}{3}}$. **32.** 0.

33. $-\dfrac{\cos x}{(1+\sin x)^2}$. **34.** $-\dfrac{4\sin x}{(3-4\cos x)^2}$. **35.** $-\dfrac{1}{(\cos x+\sin x)^2}$.

36. $\sec x(\sec x+\tan x)$. **37.** $-2\cot x\,\operatorname{cosec}^2 x$.

43. (1) $\tan\dfrac{x}{2}\sec^2\dfrac{x}{2}$; (2) $\tan\left(\dfrac{x}{2}-\dfrac{\pi}{4}\right)\sec^2\left(\dfrac{x}{2}-\dfrac{\pi}{4}\right)$.

44. $\dfrac{1}{2}[(m+n)\cos(m+n)x+(m-n)\cos(m-n)x]$. **45.** $\sin\left(\dfrac{n\pi}{2}+x\right)$.

46. (1) $\cos(\pi+x)$; (2) $\cos\left(\dfrac{3\pi}{2}+x\right)$; (3) $\cos(2\pi+x)$; (4) $\cos\left(\dfrac{n\pi}{2}+x\right)$.

Examples III. c. (p. 40.)

1. $2u\dfrac{du}{dx}$. **2.** $6(x+1)^5$. **3.** $15(3x+1)^4$. **4.** $-\dfrac{1}{(1+x)^2}$.

5. $\dfrac{1}{(1-x)^2}$. **6.** $-3(1-x)^2$. **7.** $-3(x-1)^{-4}$. **8.** $-\dfrac{2}{(1+x)^3}$.

9. $\sin 2x$. **10.** $2\cos 2x$. **11.** $k\sec^2 kx$.

12. $-m\operatorname{cosec} mx\cot mx$. **13.** $\dfrac{1}{a}\sec^2\dfrac{x}{a}$. **14.** $2\cos(2x+3)$.

15. $4(1-x)^{-5}$. **16.** $2\sin 4x$. **17.** $3(2ax+b)(ax^2+bx+c)^2$.

18. $-\dfrac{2ax+b}{(ax^2+bx+c)^2}$. **19.** $-\dfrac{5}{3}(3-x)^{\frac{2}{3}}$. **20.** $7(1+3x)^{\frac{4}{3}}$.

21. $6(1-x)^{-7}$. **22.** $-\dfrac{\sin x}{(1-\cos x)^2}$. **23.** $-\dfrac{2\sin 2x}{(1-\cos 2x)^2}$.

24. $-\dfrac{\sec^2 x}{(1+\tan x)^2}$. **25.** $-\dfrac{2}{(1+2x)^2}$. **26.** $-\dfrac{5}{(5x-1)^2}$. **27.** $\dfrac{6}{(1-6x)^2}$.

Examples III. d. (p. 42.)

1. (1) 96 sq. in.; (2) $4\cdot2\%$. **2.** $a\cos\theta$. **3.** $\dfrac{\Delta A\cdot b\cos A}{\sin B}$.

4. $\dfrac{\Delta C\cdot ab\sin C}{c}$. **5.** $2\pi a\,\Delta a$. **6.** $\dfrac{\Delta c\cdot ab\cos C}{2}$.

8. $\cdot0011\%$. **9.** $0\cdot012$. **10.** $4\cdot7$ feet; 94 ft. per sec.

Examples IV. (p. 47.)

1. 75 ft. per sec., 30 ft. sec. units.　　3. $10\cdot2$ ft. sec. units.

4. 20 sq. in. per sec.

5. (1) 6 ft. per sec.; (2) 29 ft.; (3) -2 ft. sec. units.

6. (1) $-a\omega\sin(\omega t)$; (2) $a\omega\cdot0$.　　7. $b+2ct+3t^2$.　　8. 8.

9. (1) $\dfrac{\pi x}{5}$ sq. in. per min. ; (2) $2\cdot5$ sq. in. per min.

10. $6\cdot8$ in. per min. approx.　　12. $75\cdot4$ sq. in. per sec. approx.

13. $\dfrac{5}{2\pi x}$ ft. per sec., 6 in. per sec.　　14. $0\cdot349$ c. ft. per sec.

15. $4(x+20)(x+40)$ sq. ft., $2\cdot2$ sq. ft. per hour.

17. $\dfrac{400}{x^2}$ in. per sec., $6\cdot25$ in. per sec. 18. $-\dfrac{10000}{p^2}$ c. ft. per lb. pressure.

19. $4+32t$, 68. The slope gives the velocity of the body, and the slope of the chord gives the average velocity of the body between the points where the chord cuts the curve.

20. $2\cdot5$.　　21. 1.

Examples V. (p. 52.)

1. $\dfrac{32\pi a^3}{81}$.　　2. $\dfrac{4\sqrt{3}\pi a^3}{9}$.　　3. $2a^3$.　　4. $\dfrac{2\sqrt{3}\pi l^3}{27}$.

5. $1\frac{1}{3}$ in.　　7. $2\sqrt{2}a$.　　8. $49\frac{1}{2}$.　　9. $\sqrt{a^2+b^2}$.

10. $\dfrac{ah}{2(h-a)}$.　　11. $\dfrac{4\pi a^2h}{27}$.　　12. $\dfrac{a^2}{2}$.

14. Half-way down the cistern.　　15. $2\cdot546$ c. ft.　　16. 2 miles.

17. $2a_1a_2$.　　19. 432.　　20. 307 c. in.

Examples VI. (p. 60.)

7. $\dfrac{2}{\sqrt{1-4x^2}}$.　　8. $\dfrac{2}{1+4x^2}$.　　9. $-\dfrac{1}{\sqrt{4-x^2}}$.　　10. $-\dfrac{1}{\sqrt{1-x^2}}$.

11. $-\dfrac{1}{\sqrt{2x-x^2}}$.　　12. $\dfrac{12}{9x^2+12x+20}$.　　13. $\dfrac{1}{\sqrt{2ax-x^2}}$.

14. $-\cos x-\sin x$.　　15. $\dfrac{1}{\sqrt{1-x^2}}$.　　16. $-\dfrac{2}{1+x^2}$.　　17. $-\dfrac{1}{x\sqrt{x^2-1}}$.

18. $\dfrac{1}{1+x^2}$.　　19. $\dfrac{-2a}{x^2-a^2}$.　　20. $-\tan x$.　　21. $1+\log x$.

22. $\dfrac{4a^2x}{x^4-a^4}$.　　23. $\dfrac{1}{1+x^2}$.　　24. $\dfrac{\sec^2(\log x)}{x}$.　　25. $e^{ax}(a\sin bx+b\cos bx)$.

26. $e^x(1 - 4x^3 - x^4)$. **27.** $e^x - e^{-x}$. **28.** $\dfrac{1}{\sqrt{a^2 - x^2}}$. **29.** $-\dfrac{2}{\sqrt{1 - x^2}}$.

30. $\dfrac{1}{\sqrt{a^2 + x^2}}$. **31.** $\dfrac{e^x - e^{-x}}{e^x + e^{-x}}$. **32.** $\dfrac{1}{c}\left(e^{\frac{x}{c}} - e^{-\frac{x}{c}}\right)$.

33. $\dfrac{1}{2}\tan\dfrac{x}{4}\sec^2\dfrac{x}{4}$. **34.** $\sec^2\left(x - \dfrac{\pi}{4}\right)$. **35.** $-\dfrac{1}{2}\operatorname{cosec}^2\dfrac{x}{2}$.

36. $\dfrac{1}{2}[(m+n)\sin(m+n)x - (m-n)\sin(m-n)x]$.

37. $6ax\sin^2(ax^2 + b)\cos(ax^2 + b)$. **38.** $\dfrac{\cos(\log x)}{x}$.

39. $\dfrac{\sqrt{ax(x - 3a)}}{\sqrt{x - 4a}}\left[\dfrac{1}{x} + \dfrac{1}{x - 3a} - \dfrac{1}{x - 4a}\right]$. **40.** $\dfrac{2}{1 + x^2}$. **41.** $x^x(1 + \log x)$.

42. $n\left(\dfrac{x}{n}\right)^{nx}\left(1 + \log\dfrac{x}{n}\right)$. **43.** $e^{x^x} \cdot e^x$. **44.** $\dfrac{1}{\sqrt{1 + x^2}}$.

45. $\cos^{-1}x$. **46.** $\dfrac{4}{x^2 + 4}$. **47.** $-\dfrac{1}{\sqrt{1 - x^2}}$.

48. $2x\tan^{-1}x$. **49.** $2\operatorname{cosec} x$. **50.** $\dfrac{1}{2\sqrt{x^2 - 1}}$.

51. $-2\sec x$. **52.** $\sqrt{a^2 - x^2}$. **53.** $\sqrt{a^2 + x^2}$.

54. $6(a - b)\sin 2x\,(a\sin^2x + b\cos^2x)^2$. **55.** $\dfrac{\sin^3 x}{\cos^2 x}(3\cot x + 2\tan x)$

56. $\dfrac{(x - 1)^{\frac{2}{3}}}{3(x + 1)^{\frac{1}{3}}(x - 2)^{\frac{1}{2}}}\left[\dfrac{2}{x - 1} - \dfrac{1}{x + 1} - \dfrac{1}{x - 2}\right] = \dfrac{x - 5}{3(x - 1)^{\frac{1}{3}}(x + 1)^{\frac{4}{3}}(x - 2)^{\frac{3}{2}}}$.

57. $\dfrac{1}{x\log x}$. **58.** $\dfrac{1}{2}$. **60.** $\cos\theta = 1 - \dfrac{\theta^2}{\lfloor 2} + \dfrac{\theta^4}{\lfloor 4} - \dfrac{\theta^6}{\lfloor 6} \cdots$.

62. $a^n\sin\left(\dfrac{n\pi}{2} + ax\right)$. **64.** $a^n\cos\left(\dfrac{n\pi}{2} + ax\right)$.

65. $a^x(\log_e a)^n$. **66.** $a^n e^{ax}$.

Examples VII.   (p. 65.)

1. x^3. **2.** x^5. **3.** ax. **4.** $2x^3$. **5.** $3x^4$.

6. $-x^6$. **7.** $2x^7$. **8.** $x^3 - x^2$. **9.** $\dfrac{ax^3}{3}$. **10.** bx^7.

11. $\dfrac{4cx^5}{5}$. **12.** $\dfrac{x^2}{4}$. **13.** $\dfrac{x^3}{15}$. **14.** $\dfrac{x^4}{4} + \dfrac{x^3}{3}$. **15.** $\dfrac{x^3}{3} - \dfrac{ax^2}{2}$.

16. $\dfrac{x^2}{2} + x$. **17.** $x - \dfrac{x^2}{2}$. **18.** $x^2 - x$. **19.** $4x - x^2$. **20.** $\dfrac{x^4}{4} - x$.

21. $\dfrac{x^3}{3} - a^2x.$ 22. $-\dfrac{x^{-3}}{3}.$ 23. $-\dfrac{1}{x}.$ 24. $-\dfrac{x^{-4}}{4}.$ 25. $-\dfrac{x^{-16}}{10}.$

26. $\dfrac{x^{-5}}{5}.$ 27. $\log_e x.$ 28. $\dfrac{ax^3}{3} + \dfrac{bx^2}{2} + cx.$ 29. $\dfrac{5x^{-3}}{3}.$

30. $\dfrac{x^3}{3} - \dfrac{x^2}{2}.$ 31. $\dfrac{x^2}{4} + \dfrac{ax}{2}.$ 32. $\dfrac{x^3}{3} + x - \dfrac{1}{x}.$ 33. $\dfrac{4x^{\frac{5}{4}}}{25}.$

34. $\dfrac{3x^{\frac{7}{3}}}{7}.$ 35. $\dfrac{2x^{\frac{3}{2}}}{3}.$ 36. $2\sqrt{x}$ 37. $x + \log_e x.$

38. $x - \dfrac{1}{x}.$ 39. $\dfrac{x^3}{3} + \log_e x.$ 40. $-2x^{-\frac{1}{2}}.$ 41. $\dfrac{ax^2}{2} + bx + c \log_e x$

42. $\dfrac{(x+2)^4}{4}.$ 43. $\dfrac{(x-1)^5}{5}.$ 44. $\log(x-3).$ 45. $\dfrac{(x+3)^6}{6}.$

46. $-\dfrac{1}{x-1}.$ 47. $-\dfrac{(x+2)^{-6}}{6}.$ 48. $\dfrac{(x+7)^6}{6}.$ 49. $\dfrac{(x-1)^9}{9}.$

50. $\dfrac{(ax+b)^5}{5a}.$ 51. $\dfrac{(2x-3)^6}{12}.$ 52. $-\dfrac{(1-x)^5}{8}.$ 53. $-\dfrac{1}{2(2x-3)}.$

54. $-\dfrac{(1-2x)^6}{12}.$ 55. $\log(1+x).$ 56. $\dfrac{2}{3}(x-3)^{\frac{3}{2}}.$ 57. $-\dfrac{(ax+b)^{-2}}{2a}.$

58. $\dfrac{\log(ax+b)}{a}.$ 59. $-\dfrac{1}{x+2}.$ 60. $\dfrac{2(ax+b)^{\frac{3}{2}}}{3a}.$ 61. $\dfrac{\log(2x+3)}{2}.$

62. $2\sqrt{x-4}.$ 63. $\dfrac{1}{3-x}.$ 64. $-\log(3-x).$ 65. $-\dfrac{\log(a-bx)}{b}.$

66. $-\dfrac{(1-x)^4}{4}.$ 67. $-\dfrac{(1-2x)^8}{16}.$ 68. $-\dfrac{(3-x)^5}{5}.$ 69. $\dfrac{1}{2(3-2x)}.$

70. $-\dfrac{(3-4x)^4}{16}.$ 71. $x + b \log x - \dfrac{c}{x}.$ 72. $-\dfrac{1}{x} - \dfrac{1}{2x^2} + \dfrac{2}{3x^3}.$

73. $\dfrac{x^3}{3} - \dfrac{x^2}{2} - 2x.$ 74. $\dfrac{x^4}{4} - x^2 + 4x.$ 75. $\dfrac{x^5}{5} - \dfrac{3x^4}{4} + x^3 - \dfrac{x^2}{2}.$

76. $x + 2 \log(x-1).$ 77. $3x + 7 \log(x-2).$ 78. $\dfrac{x^2}{2} + x - \log(x-2).$

79. $-\dfrac{5}{x}.$ 80. $-\dfrac{1}{1+x}.$ 81. $\dfrac{1}{1-x}.$ 82. $-2 \cos \dfrac{x}{2}.$

83. $-\cos\left(\dfrac{\pi}{2}+x\right) = \sin x.$ 84. $2 \tan \dfrac{x}{2}.$ 85. $2 \sec \dfrac{x}{2}.$

86. $-\sin\left(\dfrac{\pi}{2}-x\right) = -\cos x.$ 87. $\dfrac{1}{2} \tan 2x.$ 88. $-\dfrac{\cos 2x}{2}.$

89. $-\dfrac{1}{2}\sin\left(\dfrac{\pi}{2}-2x\right) = -\dfrac{1}{2}\cos 2x.$ 90. $\dfrac{1}{2}\cos(\pi-2x) = -\dfrac{1}{2}\cos 2x.$

91. $-\dfrac{\cos ax}{a}.$ 92. $\dfrac{\sin bx}{b}.$ 93. $\dfrac{\sec ax}{a}.$ 94. $-\cos(x-a).$

95. $-\sin(a-x)$. **96.** $-\tan(a-x)$. **97.** $-\dfrac{\cot(a+bx)}{b}$.

98. $-\dfrac{\cos 6x}{6}-\dfrac{\cos 2x}{2}$. **99.** $\dfrac{1}{4}\left(\sin 2x-\dfrac{\sin 4x}{2}\right)$.

100. $-\dfrac{1}{2}\left(\dfrac{\cos(m+n)x}{m+n}+\dfrac{\cos(m-n)x}{m-n}\right)$. **101.** $\dfrac{1}{2}\left(x-\dfrac{\sin 2x}{2}\right)$.

102. $\dfrac{1}{2}\left(x+\dfrac{\sin 2x}{2}\right)$. **103.** $\dfrac{1}{2}\left(x+\dfrac{\sin 2ax}{2a}\right)$. **104.** $\dfrac{1}{2}\left(x-\dfrac{\sin 2ax}{2a}\right)$.

105. $\dfrac{1}{2}\left[x-\dfrac{\sin 2(ax+b)}{2a}\right]$. **106.** $\dfrac{1}{2}\left[x+\dfrac{\sin 2(bx+c)}{2b}\right]$.

107. $-e^{-x}$. **108.** $ae^{\frac{x}{a}}$. **109.** e^x-e^{-x}. **110.** $a\left(e^{\frac{x}{a}}-e^{-\frac{x}{a}}\right)$.

111. $\dfrac{10^x}{\log_e 10}$. **112.** $-\dfrac{1}{2}e^{-2x}$. **113.** e^x-e^{-x}. **114.** $\dfrac{1}{2}(e^{2x}+4x-e^{-2x})$.

Examples VIII. (p. 73.)

1. $5dx$. **2.** $2x\,.\,dx$. **3.** $4x\,.\,dx$. **4.** $\dfrac{dx}{2y}=\dfrac{dx}{2\sqrt{x}}$.

5. $2\cos 2x\,.\,dx$. **6.** $2\sin x\cos x\,.\,dx$. **7.** $2\tan x\sec^2 x\,.\,dx$

8. $\sec^2\dfrac{x}{2}\tan\dfrac{x}{2}\,.\,dx$. **9.** $\dfrac{dx}{2\sqrt{1+x}}$ or $\dfrac{dx}{2y}$. **11.** $\dfrac{u-x}{u}\,.\,du$.

12. $\dfrac{(3x-4)^3}{24}$. **13.** $\dfrac{(x^3-3)^6}{5}$. **14.** $\log(ax^2+bx+c)$

15. $\dfrac{(2x+3)^6}{16}$. **16.** $-\dfrac{(3-4x)^4}{16}$. **17.** $-\dfrac{(1-2x)^{\frac{3}{2}}}{3}$.

18. $\dfrac{(x^2+1)^4}{8}$. **19.** $\dfrac{(x^3-1)^3}{9}$. **20.** $\dfrac{(x^n+a^n)^4}{4n}$.

21. $-\sqrt{1-2x}$. **22.** $-\dfrac{1}{2}\log(1-x^2)$. **23.** $\dfrac{\log(1+3x^2)}{6}$.

24. $\dfrac{1}{4}\log(4x-3)$. **25.** $-\log(\cos x)$. **26.** $\log(\sin x)$.

27. $-\dfrac{1}{2}\cos 2x$. **28.** $-2\cos\dfrac{x}{2}$ **29.** $\dfrac{1}{6}\tan 6x$.

30. $3\tan\dfrac{x}{3}$. **31.** $-\dfrac{1}{3}\cot 3x$. **32.** $-\sqrt{1-x^3}$.

33. $2\sqrt{ax^2+bx+c}$. **34.** $2\sqrt{x^2-5x+3}$. **35.** $2\sqrt{x^3+x-2}$.

36. $\dfrac{1}{2}\log(1+x^3)$. **37.** $\dfrac{(ax^2+b)^{\frac{3}{2}}}{3a}$. **38.** $\dfrac{(x^2-4x-3)^6}{12}$.

39. $-\dfrac{\log(3+5\cos x)}{5}$. **40.** $\dfrac{-1}{\sqrt{x^2-1}}$ **41.** $\sin x-\dfrac{\sin^3 x}{3}$

42. $\dfrac{\sin 2x}{2} - \dfrac{\sin^3 2x}{6}$.   43. $\dfrac{1}{5}(x^3 + 2ax + b)^{\frac{3}{2}}$.   44. $\dfrac{\log(ax^4 + b)}{4a}$.

45. $-\dfrac{1}{2}\log(\cos 2x)$.   46. $\dfrac{(x^4 - 1)^{\frac{3}{2}}}{6}$.   47. $\dfrac{1}{2}\sin^{-1}x + \dfrac{x}{2}\sqrt{1 - x^2}$.

48. $\log(x + \sqrt{1 + x^2})$.   49. $\log(x + \sqrt{x^2 - 1})$.

50. $\dfrac{2}{27}\left[\dfrac{u^7}{7} + \dfrac{2u^5}{5} + \dfrac{u^3}{3}\right]$, where $u = \sqrt{3x - 1}$.   51. $\dfrac{1}{2}(x - \sin x)$.

52. $2\sqrt{\sin x}$.   53. $e^{\tan x}$.   54. $\dfrac{1}{2}\tan^{-1}\dfrac{x}{2}$.

55. $\dfrac{1}{3}\tan^{-1}3x$.   56. $\dfrac{3}{2}\log(1 + 2x) - x$.   57. $\dfrac{1}{2}\log(x^2 + 2x + 2)$

58. $\dfrac{(x^5 + 1)^8}{40}$.   59. $\sqrt{x^2 + 1}$.   60. $\dfrac{x}{\sqrt{1 + x^2}}$.   61. $\dfrac{x}{a^2\sqrt{x^2 + a^2}}$.

62. $\log(x + \sqrt{x^2 + 4})$.   63. $\dfrac{9}{2}\sin^{-1}\dfrac{x}{3} + \dfrac{x}{2}\sqrt{9 - x^2}$.

64. $\log(x + \sqrt{x^2 - 4})$.   65. $\dfrac{1}{2}\tan^{-1}\left(\dfrac{\sin x}{2}\right)$.

66. $\dfrac{1}{2}\tan^{-1}\left(\dfrac{\tan x}{2}\right)$.   67. $\dfrac{2}{75}\sqrt{5x + 4}\,(5x - 8)$.

68. $\dfrac{2}{15}\sqrt{x + 1}\,(3x^2 - 4x + 8)$.   69. $\dfrac{2}{3}\sqrt{x + 2}\,(x - 4)$.

70. $\dfrac{1}{2}\log\left(\dfrac{1 + x}{1 - x}\right)$.   71. $x + \log(x - 1)$.   72. $\log\left(\dfrac{x - 3}{x - 2}\right)$.

73. $\dfrac{1}{2}\log\left(\dfrac{x - 1}{x + 1}\right)$.   74. $-\dfrac{1}{8}\log(1 - 4x^2)$.   75. $\dfrac{1}{4}\log\left(\dfrac{x - 1}{x + 3}\right)$.

76. $\log\left(\dfrac{3x - 1}{2x + 1}\right)$.   77. $\dfrac{1}{a - b}\log\left(\dfrac{x - a}{x - b}\right)$.   78. $\log(x - 1)(x - 2)(x - 3)$.

79. $\log\dfrac{(x - 1)(x - 2)}{(x + 1)}$.   80. $\log\dfrac{x^2 - 1}{x^2}$.   81. $-\tan^{-1}x - \dfrac{1}{x}$.

82. $-2\tan^{-1}2x - \dfrac{1}{x}$.   83. $\dfrac{5}{2}\log(x^2 + 4) + \dfrac{1}{2}\tan^{-1}\dfrac{x}{2}$.

84. $\dfrac{x^{\frac{1}{2}}}{5}(x^2 - 25)$.   85. $\dfrac{2}{3}(2a + x)^{\frac{3}{2}}$.   86. $\log\left(\dfrac{x - 2}{x - 1}\right)$.

87. $\dfrac{x^3}{6} + \dfrac{a^3}{2}\log x$.   88. $\dfrac{1}{\sqrt{3}}\tan^{-1}\dfrac{x}{\sqrt{3}}$.   89. $\dfrac{1}{2}\sec^{-1}\dfrac{x}{2}$.

90. $\dfrac{3}{2}\log(1 + x^2) + \tan^{-1}x$.   91. $\sin^{-1}x - 2\sqrt{1 - x^2}$.

92. $\dfrac{1}{\sqrt{2}}\tan^{-1}\dfrac{x}{\sqrt{2}} + \dfrac{1}{2}\log(2 + x^2)$.   93. $\dfrac{1}{2}\tan 2x - x$.

94. $2\cos\dfrac{x}{2}\left(\dfrac{1}{3}\cos^2\dfrac{x}{2} - 1\right)$.   95. $\dfrac{1}{1 - \sin x}$.

Examples IX. (p. 77.)

1. $7\frac{1}{2}$. 2. 32. 3. $7\frac{1}{2}$. 4. $4\frac{1}{2}$. 5. $\log_e 3 = 1 \cdot 10$.

6. $\frac{1}{3}$. 7. 4. 8. 18. 9. $1\frac{1}{3}$. 10. 2.

11. $\frac{\pi}{4} = \cdot 7854$. 12. $\frac{a^3}{3}$. 13. $60\frac{2}{4}$. 14. $\frac{1}{2}$. 15. 4.

16. $20\frac{1}{3}$. 17. 2. 18. $\frac{\pi}{4} = \cdot 7854$. 19. $\pi = 3 \cdot 1416$.

20. $\frac{1}{16}$. 21. $\frac{2}{3}$. 22. 4. 23. $\frac{\pi}{8} = \cdot 3927$.

24. $\frac{1}{2}(\pi - 2) = \cdot 5708$. 25. $6 - 2\log_e 7 = 2 \cdot 108$. 26. $\frac{\pi a^4}{16}$.

Examples X. b. (p. 83.)

1. 21 ft. 2. $64\frac{1}{3}$ ft. 3. 27 ft. per sec.; 90 ft. 4. $4\sqrt{21} = 18 \cdot 33$ ft.

5. $v = \frac{at^2}{2} + bt + u$; $s = \frac{at^3}{6} + \frac{bt^2}{2} + ut + c$. 6. $\frac{ds}{dt} = ut$; $s = \frac{ut^2}{2} + a$.

7. $x = u \cos at$; $y = u \sin at - \frac{1}{2}gt^2$. 8. $x = ft$; $y = \frac{ft}{2}(t+2)$.

10. $x = \frac{1}{4}(2\sqrt{a} - t)^2$. 11. 65 ft. per sec.

Examples XI. (p. 86.)

1. $x(\log x - 1)$. 2. $e^x(x - 1)$. 3. $e^x(x^2 - 2x + 2)$.

4. $\frac{1}{a^2}\sin ax - \frac{x}{a}\cos ax$. 5. $\frac{x}{a}\sin ax + \frac{1}{a^2}\cos ax$.

6. $2\cos x + 2x \sin x - x^2 \cos x$. 7. $\frac{x}{2}\sqrt{x^2 - a^2} - \frac{a^2}{2}\log(x + \sqrt{x^2 - a^2})$.

8. $\frac{x}{2}\sqrt{x^2 + 1} + \frac{1}{2}\log(x + \sqrt{x^2 + 1})$. 9. $\frac{x}{2}\sqrt{x^2 + 25} + \frac{25}{2}\log(x + \sqrt{x^2 + 25})$.

10. $x \log_{10}\left(\frac{x}{e}\right)$. 11. $x \log(x + \sqrt{a^2 + x^2}) - \sqrt{a^2 + x^2}$.

12. $\frac{1}{4}\cos 2x + \frac{x}{2}\sin 2x - \frac{x^2}{2}\cos 2x$.

13. (1) $-\frac{\cos^2 x}{2}$; (2) $-\frac{\cos 2x}{4} = \frac{1 - 2\cos^2 x}{4}$, which differs from the first result by a constant.

14. $\frac{x}{a}e^{ax} - \frac{1}{a^2}e^{ax}$. 15. $\frac{x\sqrt{a^2 - x^2}}{2} + \frac{a^2}{2}\sin^{-1}\frac{x}{a}$. 16. $\frac{x^6}{6}\left(\log x - \frac{1}{6}\right)$.

17. $\frac{1}{2}[(1 + x^2)\tan^{-1}x - x]$. 18. $x \tan^{-1}x - \frac{1}{2}\log(1 + x^2)$.

Examples XII. a. (p. 95.)

2. $\frac{2k}{3}(h-2a)$.   **3.** $\frac{h}{3}(k+4a)$.   **4.** πab.   **5.** $1\frac{1}{3}$.

6. $\frac{1}{6}$.   **7.** $\frac{256\sqrt{2}}{3}=120\cdot68$.   **8.** 9.   **9.** $11\frac{1}{4}$.

10. 72.   **11.** $\frac{a^2\theta}{2}$.   **12.** πab.   **13.** $15\frac{3}{4}$.

14. 2.   **15.** $114\cdot59$.   **16.** $\frac{b^4-a^4}{4c^2}$.   **17.** $1\frac{1}{3}$.

18. $4\pi=12\cdot57$.   **19.** $5\pi=15\cdot71$.   **20.** $20-9\log_e3=10\cdot11$.

Examples XII. b. (p. 98.).

1. a^2.   **2.** $\frac{3\pi a^2}{2}$.   **3.** $\frac{a^2}{4}(\pi+2)$.   **4.** $24\pi=75\,4$

5. $\frac{ab}{2}\tan^{-1}\left(\frac{a\tan\theta}{b}\right)$.

Examples XII. c. (p. 103.)

1. $\pi(\sqrt{3}-1)=2\cdot3$ sq. ft.   **2.** $8\pi(2+\sqrt{3})=93\cdot8$ sq. in.

3. $4a$.   **4.** $8a$.

Examples XII. d. (p. 104.)

1. $\frac{\pi a^2h}{3}$.   **2.** $\frac{4}{3}\pi a^3$.   **3.** $\frac{\pi h^2}{3}(3a-h)$.   **4.** $2\pi ah^2=\frac{\pi hk^4}{2}$.

5. $26\pi=81\cdot68$.   **6.** $\frac{40\pi}{3}=41\cdot89$.   **7.** $\frac{\pi}{4}(e^{16}-1)$.   **8.** $2\pi ah^2$.

Examples XIII. a. (p. 109.)

1. At a distance $\frac{4a}{3\pi}$ from the centre along the radius bisecting the area.

2. At equal distances $\frac{4a}{3\pi}$ from the bounding radii.

3. If h is the height of the cone, the c.g. is on the axis and at a distance $\frac{h}{4}$ from the base of the cone.

4. At a distance $a\frac{\sin a}{a}$ from the centre along the radius bisecting the arc.

5. At a distance $\dfrac{3h}{5}$ from the vertex along the axis.

6. Referred to the principal axes of the ellipse the c.g. is at the point $\left(\dfrac{4a}{3\pi}, \dfrac{4b}{3\pi}\right)$.

7. At a distance $\dfrac{4a}{3\pi}$ or $\dfrac{4b}{3\pi}$ along the axis bisecting the area.

8. The c.g. is at the point $\left(\dfrac{3h}{5}, \dfrac{3k}{8}\right)$.

9. The area $= 20 \log_e 3 = 21\cdot 97$.

The c.g. is at the point $\left(\dfrac{8}{\log_e 3}, \dfrac{5}{3\log_e 3}\right)$, i.e. $(7\cdot 4,\ 1\cdot 5)$.

10. (1) $31\cdot 5$ sq. in. ; (2) $5\cdot 9$ c. in.

Examples XIII. b. (p. 111.)

1. (1) The total pressure $= \dfrac{2}{3}\rho \times$ area $\times$ altitude of the $\triangle$.

(2) The centre of pressure is at a depth of $\dfrac{3}{4}$ of the altitude.

2. (1) The total pressure $= \dfrac{\rho a k}{2}\left(h + \dfrac{2k}{3}\right)$, where $k =$ alt. of $\triangle$, and $a =$ its base.

(2) The centre of pressure is at a depth $\dfrac{k}{2}\left(\dfrac{4h + 3k}{3h + 2k}\right) + h$.

3. (1) The total pressure $= \dfrac{\rho a b^2}{2}$.

(2) The centre of pressure is at a depth $\dfrac{2b}{3}$.

4. (1) The total pressure $= \rho a b\left(h + \dfrac{b}{2}\right)$.

(2) The centre of pressure is at a depth $\dfrac{b}{3}\left(\dfrac{3h + 2b}{2h + b}\right) + h$.

5. (1) The total pressure $= \dfrac{\rho a k^2}{6}$, where $k =$ alt. of $\triangle$.

(2) The centre of pressure is at a depth $\dfrac{k}{2}$.

6. (1) The total pressure $= \dfrac{\rho a k}{2}\left(h + \dfrac{k}{3}\right)$.

(2) The centre of pressure is at a depth $\dfrac{k}{2}\left(\dfrac{2h + k}{3h + k}\right) + h$.

7. (1) The total pressure $= \dfrac{\rho a b^2}{2} \sin a$.

(2) The distance of the centre of pressure from the upper side $= \dfrac{2b}{3}$

8. (1) The total pressure $=\dfrac{\rho ab}{2}(2k + b\sin a)$.

(2) The distance of the centre of pressure from the upper side
$$=\frac{b}{3}\left[\frac{3k + 2b\sin a}{2k + b\sin a}\right].$$

9. (1) The total pressure $=\dfrac{\rho \sin \theta}{2}=31\cdot15\sin\theta$.

(2) The distance of the centre of pressure from the upper edge
$$=8 \text{ inches.}$$

(3) The required angle $=\tan^{-1}0\cdot475 = 25^\circ\ 24\tfrac{1}{2}'$.

10. (1) $\dfrac{2}{3}\dfrac{x^2+xy+y^2}{x+y}$; (2) $\dfrac{a}{2}(y^2 - x^2)$ grams wt.

Examples XIII. c. (p. 113.)

1. $6\tfrac{2}{3}$ ft.-lb. **2.** 50 ft.-lb. **3.** $144\times180\log_e 4 = 35900$ ft.-lb

Examples XIV. (p. 119.)

1. $\dfrac{Ma^2\omega^2}{4}$. **2.** $\dfrac{mb^2}{3}$, $\dfrac{mb^2\omega^2}{6}$. **3.** $\dfrac{Ma^2}{4}$. **4.** $\dfrac{mb^2}{12}$.

5. $Ma\left(\dfrac{a+3b}{3}\right)$. **6.** $\dfrac{mh^2}{2}$. **7.** $\dfrac{2a^2M}{5}$.

8. $\dfrac{200}{3}$ cm.; $1\cdot64$ secs. **9.** $\dfrac{5Mr^2}{4}$; $\pi\sqrt{\dfrac{5r}{g}}$. **10.** $\dfrac{3Ma^2}{2}$; $\pi\sqrt{\dfrac{6r}{g}}$.

Examples XV. (p. 121.)

1. $(0, a)$. **2.** $\dfrac{1}{2}\left(e^{\frac{x}{a}} - e^{-\frac{x}{a}}\right)$.

Examples XVI. (p. 130.)

1. $v = -na\sin(nt + m)$; $f = -n^2x$.

3. (i) $2\sqrt{9 - x^2}$ ft. per sec.; (ii) $2\sqrt{5}=4\cdot47$ ft. per sec.;

(iii) $4\sqrt{2}=5\cdot66$ ft. per sec.; (iv) 6 ft. per sec.

(v) π.

4. C moves with Simple Harmonic Motion. If the $\angle POC = \theta$ at the beginning of motion, C rises through a distance $a(1-\cos\theta)$, and falls through a distance $a(1+\cos\theta)$ during a half-revolution.

5. The periodic time $=\dfrac{\pi AB}{u}$.

Examples XVII. (p. 135.)

1. $\dfrac{\cos x}{3}(\cos^2 x - 3)$.   2. $\cos x\left(1 - \dfrac{2}{3}\cos^2 x\right) = \dfrac{1}{2}\left(\cos x - \dfrac{1}{3}\cos 3\alpha\right)$.

3. $\dfrac{\cos^3 x}{15}(3\cos^2 x - 5)$.   4. $\dfrac{1}{\sqrt{3}}\log\left(\dfrac{\sqrt{3}+x}{\sqrt{3}-x}\right) + \dfrac{1}{2}\log(3 - x^2)$.

5. $\dfrac{1}{6}\log\left(\dfrac{1-\cos 3\theta}{1+\cos 3\theta}\right)$.   6. $\dfrac{1}{2}\log(1+\cos^2 x) - \log\cos x = \dfrac{1}{2}\log(1+\sec^2 x)$.

7. $\dfrac{1}{17}\left[3\log(2x-3) - \dfrac{3}{2}\log(x^2+2) + 2\sqrt{2}\tan^{-1}\dfrac{x}{\sqrt{2}}\right]$.   8. $-\log(\cos\overline{x+a})$.

9. $\dfrac{1}{16}[4\sin 2x - 4x - \sin 4x]$.   10. $\dfrac{4}{3}\sin^{\frac{3}{4}}x - \dfrac{4}{11}\sin^{\frac{11}{4}}x$.

11. $\dfrac{1}{12}\tan^{-1}\left(\dfrac{4}{3}\tan\theta\right)$.   12. $\sqrt{\dfrac{1+x}{1-x}}$.   13. $\dfrac{1}{2\sqrt{7}}\log\left(\dfrac{x-\sqrt{7}}{x+\sqrt{7}}\right)$.

14. $\sin^{-1}\left(\dfrac{2x-3}{4}\right)$.   15. $\log(x+2+\sqrt{x^2+4x+5})$.

16. $\dfrac{x}{2}\sqrt{x^2+9} + \dfrac{9}{2}\log(x+\sqrt{x^2+9})$.   17. $\dfrac{1}{2}\log(2x+\sqrt{4x^2+1})$.

18. $\dfrac{x}{2}\sqrt{1-4x^2} + \dfrac{1}{4}\sin^{-1}2x$.   19. $\tan^{-1}\left(\dfrac{2x-1}{3}\right)$.   20. $\dfrac{1}{3}\sin^{-1}(3x-1)$.

21. $\dfrac{1}{12}\log\left(\dfrac{2x-3}{2x+3}\right)$.   22. $\dfrac{1}{2}\log(\sec 2x + \tan 2x) = \dfrac{1}{2}\log\left[\tan\left(\dfrac{\pi}{4}+x\right)\right]$.

23. $\dfrac{1}{\sqrt{8}}\tan^{-1}\left(\dfrac{3\tan x - 1}{\sqrt{8}}\right)$.   24. $\dfrac{1}{2}\tan^{-1}\left(\dfrac{1}{2}\tan\dfrac{x}{2}\right)$.

25. $\sin^{-1}\left(\dfrac{x-2}{2}\right)$.   26. $\dfrac{1}{4}\log\left(\dfrac{x}{4-x}\right)$.

27. $\dfrac{2}{\sqrt{19}}\tan^{-1}\left(\dfrac{2x+1}{\sqrt{19}}\right)$.   28. $\log(x+1+\sqrt{x^2+2x+2})$.

29. $\dfrac{1}{2}(\log_e x)^2$.   30. $\log_e a\,\log_e x + \dfrac{1}{2}(\log x)^2$.

31. $\dfrac{1}{54}\left(\tan^{-1}\dfrac{x}{3} + \dfrac{3x}{x^2+9}\right)$.   32. $\dfrac{1}{4}\left(\log\dfrac{x+1}{x-1} - \dfrac{2x}{x^2-1}\right)$.

33. $x^3+3x^2+12x+24\log(x-2)$.   34. $-(9-x^2)^{\frac{3}{2}}(x^2+6)$.

35. $\dfrac{2}{\sqrt{17}}\tan^{-1}\left(\dfrac{\tan\dfrac{x}{2}}{\sqrt{17}}\right)$.   36. $\dfrac{1}{\sqrt{17}}\log\left(\dfrac{\sqrt{17}+\tan\dfrac{x}{2}}{\sqrt{17}-\tan\dfrac{x}{2}}\right)$.

37. $-\dfrac{\sqrt{4-x^3}}{4x}$.
38. $2\log\left(\sec\dfrac{x}{2}+\tan\dfrac{x}{2}\right)=2\log\left[\tan\left(\dfrac{\pi}{4}+\dfrac{x}{4}\right)\right]$.

39. $\dfrac{1}{2}[\sec x\tan x+\log(\sec x+\tan x)]$.

40. $\dfrac{1}{4}\log\left[\dfrac{2-\sqrt{4-x^2}}{2+\sqrt{4-x^3}}\right]=\dfrac{1}{2}\log\left(\dfrac{2-\sqrt{4-x^3}}{x}\right)$.

Revision Paper I. (p. 137.)

1. $\dfrac{4a^2x}{(x^2+a^2)^3}$. $e^{-3x}[4\sin(a-4x)-3\cos(a-4x)]$. $\sec x$.

2. (1) 30 ft. per sec., 8 ft.-sec. units;   (2) 34 ft., 8 ft. per sec.

3. $3(y-y_1)(x_1^2-1)+x-x_1=0$.

4. $-\dfrac{1}{2}\sqrt{1-4x}$; $\sin^{-1}\dfrac{x}{2}+\sqrt{4-x^3}$; $\dfrac{1}{2\sqrt7}\tan^{-1}\left(\dfrac{\sqrt7x}{2}\right)$.   **5.** 1241 s. cm

Revision Paper II. (p. 137.)

1. 24.   **2.** $-3(1-2x)^{\frac{1}{2}}$;   (1) $\dfrac{2a}{(a-x)^2}$;   (2) $\dfrac{\sqrt{x-a}-\sqrt{x+a}}{4a\sqrt{x^2-a^2}}$.

3. $2y=3x^2-10x+9$.   **4.** 193.

5. (1) $\dfrac{1}{10}\log_e\dfrac{x-7}{x+3}$;   (2) $\dfrac{1}{10}\log_e(1+5x^2)$;   (3) $\dfrac{1}{\sqrt5}\tan^{-1}(\sqrt5x)$.

Revision Paper III. (p. 138.)

1. $-(x-2)^{-2},\,2(x-2)^{-3},\ -6(x-2)^{-4},\,(-1)^n\underline{|n}(x-2)^{-n-1}$.

2. $\dfrac{b\cdot db-a\cdot da}{c}$.   **3.** (1) 20.   (2) It has no maximum value.

4. (1) $\dfrac{1}{12}(6x-\sin6x)$;   (2) $-2\sqrt{1-x^2}-3\sin^{-1}x$;   (3) $\log(x+\sqrt{16+x^2})$.

5. $13\frac{1}{3}$.

Revision Paper IV. (p. 139.)

1. $-a(ax+b)^{-2},\,2a^2(ax+b)^{-3},\ -6a^3(ax+b)^{-4},\,(-1)^n\underline{|na^n}(ax+b)^{-n-1}$.

4. $\tan^{-1}\dfrac{12}{41}=16°\,19'$.   Length intercepted$=5\frac{1}{3}$.

5. (1) $2\sqrt{x}-4\tan^{-1}\dfrac{\sqrt{x}}{2}$;   (2) $x+4\sqrt{x}+4\log(\sqrt{x}-1)$;   (3) $\dfrac{2}{3}\sqrt{x-1}(x+2)$.

Revision Paper V. (p. 139.)

2. $\dfrac{1}{(x-2)^2} - \dfrac{1}{(x-3)^2}.$

4. (1) $e^x \sin x$; (2) $\dfrac{2}{3b^2}\sqrt{a+bx}(bx-2a)$; (3) $\dfrac{2}{\sqrt{7}}\tan^{-1}\left(\dfrac{1}{\sqrt{7}}\tan\dfrac{x}{2}\right).$

5. $5\frac{1}{3}.$

Revision Paper VI. (p. 140.)

3. $\dfrac{2}{15}(x+1)^{\frac{3}{2}}(3x-2).$

Revision Paper VII. (p. 140.)

1. (1) $y\dfrac{dy}{dx}=x$; (2) $y=\sqrt{a^2-x^2}$. **2.** $\dfrac{30}{\pi x^2}$ ft. per min. **3.** $\dfrac{3\pi}{2}=4.71$

4. Maximum 400, minima 22 and -6890.

5. (1) $\dfrac{1}{2}\tan^{-1}(2\sin x)$. (2) $\dfrac{3}{2}\tan^{-1}(2x)+\dfrac{5}{8}\log(1+4x^2)$;

  (3) $\dfrac{e^{2x}}{5}(2\sin x - \cos x).$

Revision Paper VIII. (p. 141.)

1. (1) $\cos(x+a)$; (2) $-3x(1-x^2)^{\frac{1}{2}}.$

2. Maximum at the point (1, 19); minimum at (5, -13).

4. $2\frac{1}{4}.$ **5.** 33·51 sq. ft.

Revision Paper IX. (p. 141.)

(1) $\dfrac{\cdot 2ab}{a^2-b^2x^2}$; (2) $e^{x\log x}(1+\log x)$. **2.** (1) $\dfrac{1}{2}\sin^2 x$; (2) $-\dfrac{\cos 2x}{4}.$

3. (1) $-\sin^{-1}\left(\dfrac{x-a}{a}\right)$; (2) $\dfrac{1}{a}\sin^{-1}\left(\dfrac{x-a}{x}\right)$. **5.** (0, 0), (2, -4).

Revision Paper X. (p. 142.)

1. (1) $\dfrac{2}{\sqrt{1+4x^2}}$; (2) $\dfrac{x^2-1}{2x^2\sqrt{x+\dfrac{1}{x}}}$. **2.** $\dfrac{1}{2}\log(x^2+\sqrt{1+x^4}).$

3. (1) mg; (2) $mg(u+gt)$. **4.** $\pm\sqrt{2}.$

Revision Paper XI. (p. 143.)

1. (1) $b(1 + \log ax)(ax)^{bx}$; (2) $\dfrac{4}{1 + 4x^2}$.

2. (1) $\dfrac{1}{64}[\log(1 + 4x) + 4x(2x - 1)]$; (2) $172\cdot 8$.

3. 44 ft. per sec. **4.** The rate $= 2\pi r = \dfrac{2A}{r}$.

Revision Paper XII. (p. 143.)

3. $1\cdot 2$ c. ft. **4.** $4\log_e 2 = 2\cdot 77$. **5.** (1) $996\cdot 8$ lb. wt.; (2) 2 ft. 8 in

Revision Paper XIII. (p. 144.)

1. (1) $\dfrac{1}{20}$; (2) $\dfrac{\pi}{2}$. **2.** 934,500 lb. wt. $= 417\cdot 2875$ tons wt.

3. The area $= \dfrac{47}{48}$. **5.** (1) $747\cdot 6$ lb. wt.; (2) $6\frac{1}{18}$ ft.

Revision Paper XIV. (p. 144.)

1. (1) $e^x \sin 2x$; (2) $\log \tan \theta + \dfrac{1}{2}\tan^2\theta$. **3.** $0\cdot 9$ per cent.

4. $66\cdot 5$ ft. lb. **5.** $\dfrac{dT}{dt} = -k(T - a)$.

Revision Paper XV. (p. 145.)

1. (1) $\dfrac{1}{7}\left[\dfrac{1}{3}\tan^{-1}\dfrac{x}{3} - \dfrac{1}{4}\tan^{-1}\dfrac{x}{4}\right]$; (2) $-\dfrac{1}{2}[\sin^{-1}\sqrt{1 - x^2} + x\sqrt{1 - x^2}]$.

2. $1\cdot 944$ c. in. per sec. **3.** 1 sq. ft. **4.** 10660 ft. lb. approx

£4.50

Lightning Source UK Ltd.
Milton Keynes UK
UKHW042244150119
335632UK00001B/48/P

9 781447 457473

THE CALCULUS
FOR BEGINNERS

BY

W. M. BAKER, M.A.

LATE HEADMASTER OF THE MILITARY AND CIVIL DEPARTMENTS
AT CHELTENHAM COLLEGE